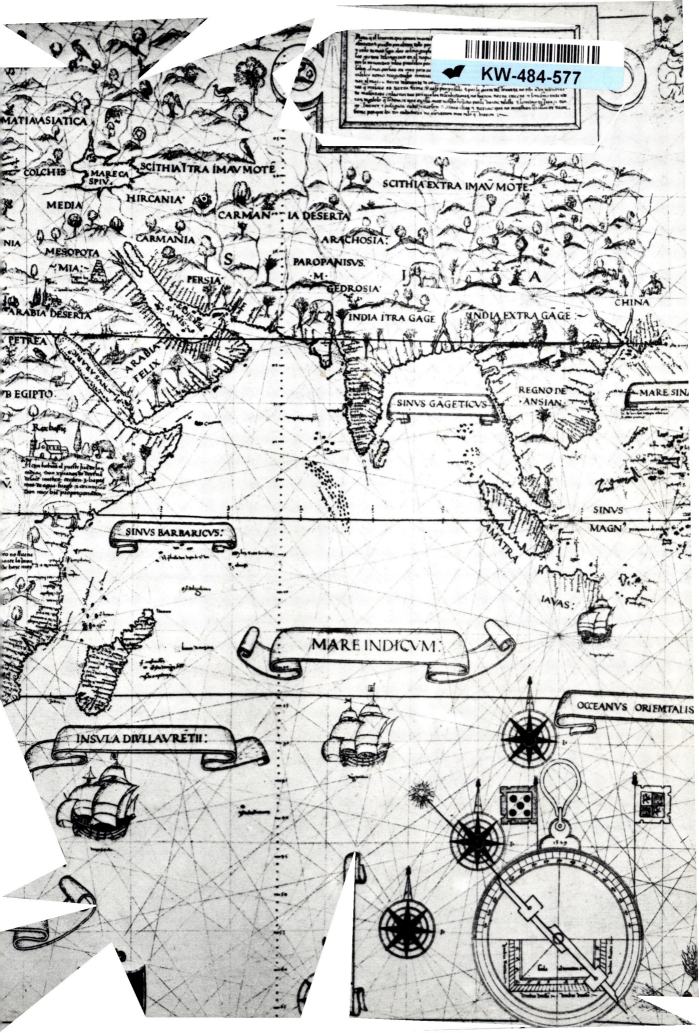

History Eye-witness

Explorers

Neil Grant

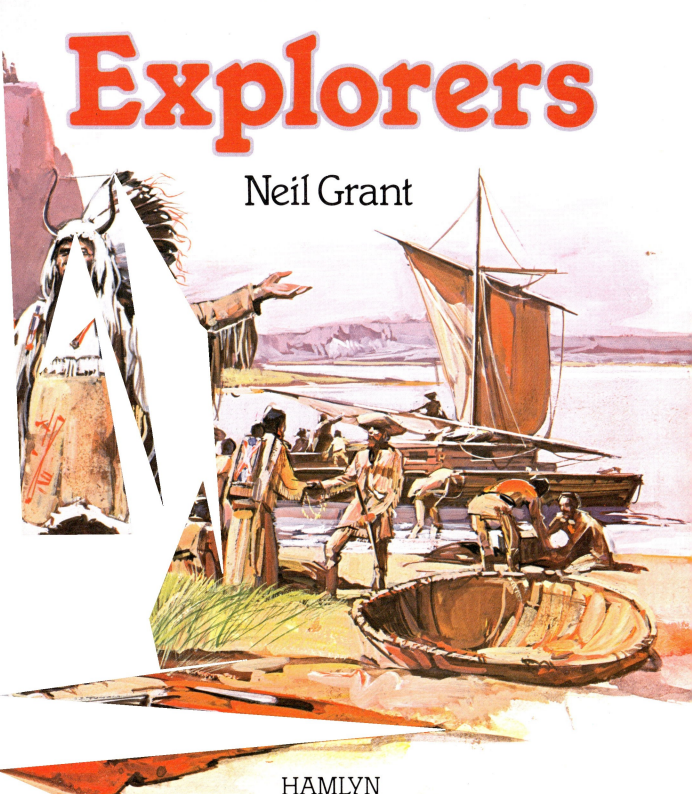

HAMLYN
London · New York · Sydney · Toronto

ACKNOWLEDGEMENTS

Aldus Archives 81 top right; H. Aschehoug & Co., Oslo 83 bottom right; British Antarctic Survey, Cambridge 89 centre, 91 bottom right; Central Office and Museum of National Antiquities, Stockholm – S. Hallgrens 14; Bruce Coleman Ltd., London 30 top; Hamlyn Group Picture Library 71 left; Mansell Collection, London 69 top right; Axel Poignant, London 53 right; Royal Geographical Society – photo © Aldus Books 73 bottom left.

Extract from BY THE GREY-GULF WATER by A. B. Paterson page 69 reprinted from THE COLLECTED VERSE OF A. B. PATERSON by permission of the Copyright Owner and the Angus & Robertson Publishers, Sydney.

First published 1979 by
The Hamlyn Publishing Group Limited
London · New York · Sydney · Toronto
Astronaut House, Feltham, Middlesex, England

Illustrations by Mike Codd, Peter Dennis and Gwen Green
Maps by Stuart Perry

ISBN 0 600 33163 6

Printed in Italy, by New Interlitho

Contents

Pytheas

A Greek visits Britain 320 BC

In the fourth century BC a rich colony of Greek merchants lived where the French port of Marseilles is now. Among them was Pytheas, a scholar as well as a merchant, who had discovered that the Pole Star does not mark the North Pole *exactly*, and who had been able to work out correctly the latitude of his home town.

The Greeks were the most intelligent and most advanced people of the time. That was not only the opinion of the Greeks themselves. Others thought so too, including the Romans, who were soon to become such a powerful nation in the ancient world. That world was a small place – or so it seems to us. It stretched no farther than the Mediterranean Sea and the countries around it.

The Greeks were not supreme in everything. The greatest navigators and traders of the Mediterranean were the Carthaginians. Only the Carthaginians sailed beyond the Pillars of Hercules, out of Mediterranean waters. To keep trade in their own hands they prevented others from following them. But the Greeks of Marseilles were eager to break into that trade, and Pytheas was the man who slipped past the Carthaginians' blockade, past the Pillars of Hercules and out into the Atlantic Ocean.

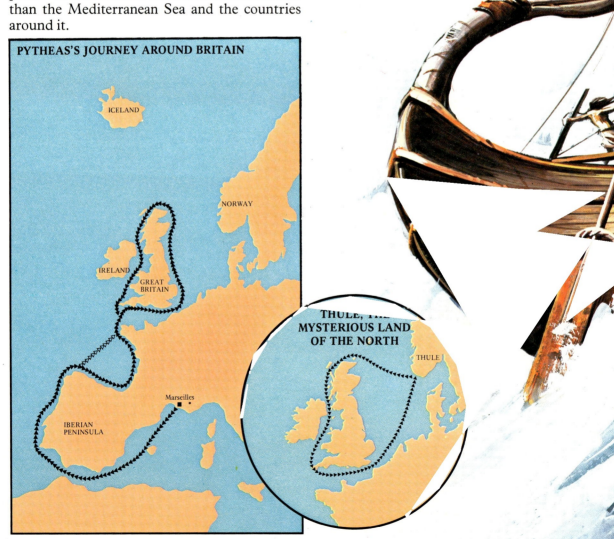

PYTHEAS'S JOURNEY AROUND BRITAIN

ICELAND

NORWAY

IRELAND

GREAT BRITAIN

Marseilles

IBERIAN PENINSULA

THULE, THE MYSTERIOUS LAND OF THE NORTH

THULE

The whole mast could be dismantled and left ashore when the ship was on a short journey.

Pytheas probably sailed with at least two ships, probably made of fir with an oak keel. They would have been powered by oars as well as by sail.

There were two large steering oars at the stern instead of a rudder.

The cold weather of northern Britain no doubt came as a shock to sailors used to a Mediterranean climate. They probably bartered for warmer clothing.

grain

oil

wine

olives

honey

oranges

9

Pytheas was away for about six years. Unfortunately, the account he wrote of his travels was lost many centuries ago. The only parts of it we know are those which were quoted by later writers. We do not know exactly what Pytheas did or where he went.

Journey to Britain

This much we do know. We know he sailed along the coasts of Spain and Portugal, around the Bay of Biscay to the Isle of Ushant, and from there across the English Channel. Tin had been mined in Cornwall for centuries before Pytheas arrived, and the area was probably an important source of tin for the whole Mediterranean region.

As tin was a very useful metal, and was needed to make bronze, Pytheas was interested to find exactly where the tin came from. He wrote a detailed description of the Cornish tin industry in that far-off time.

From Cornwall Pytheas sailed all the way around Britain – up the Channel and around

The Greeks were very interested in Celtic industry but thought the Britons 'backward'.

Kent, north to Scotland and down the west coast to Wales and the Bristol Channel. He must have seen Ireland as he sailed through the Irish Sea on his way south, but he did not mention it and landed only at places in Britain. The Celtic British, said Pytheas, were friendly and hospitable, though rather backward.

Pytheas may have exaggerated his travels in Britain. But his opinions are still very interesting because he was, so far as we know, the first 'tourist' in Britain and the first person to write an account of Britain and the British. He made some mistakes, especially when calculating distances. (He reckoned that Britain was more than twice as big as it actually is.) But that kind of mistake is easily made by an explorer who has no maps and no reliable way of measuring distances.

A voyage from the Mediterranean to the north of Britain does not sound much now. But in the fourth century BC it was amazing. Many people just did not believe Pytheas's account of his travels. He must have been an extraordinary man. He was certainly an extraordinary explorer, and his voyage was as great an achievement in its time as the voyage of Columbus in the fifteenth century or the first moon flight in the twentieth century.

Problems of early explorers

How was Pytheas equipped for his voyage to the unknown north? His ship was a surprisingly sturdy craft, a typical merchant vessel of the time, with a square sail on her single mast and another square sail on the bowsprit. The ship (and probably there was more than one) could also be propelled by oars, and she was steered by a paddle at the stern. She was rather broad and shallow for an ocean-going ship – the Greeks must have suffered sea sickness in the Bay of Biscay. She was rather slow too, although that hardly mattered. What was a drawback was that she was 'unhandy' – a sailors' term which means she could not manoeuvre easily. She also had to have the wind more or less behind her.

The open ocean came as a nasty shock to Greek sailors who were used to the Mediterranean. They had heard of tides but never experienced them. They were surprised by the strength of ocean currents, by the suddenness and ferocity of storms, and by the violent changes of wind, although they had heard of these things too. Thick ocean fog was another frightening novelty. No Mediterranean sailor had ever seen the features of the coastline

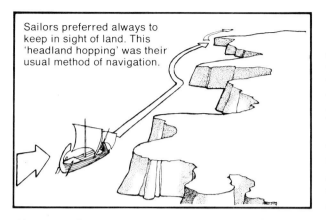

Sailors preferred always to keep in sight of land. This 'headland hopping' was their usual method of navigation.

disappear from view in a dense swirling gloom in the middle of the day. Navigating was hard enough without that. Ancient navigators followed the coast as far as possible, from one headland to another. That is quite easy in the Mediterranean, especially when you know what landmarks to look for. It is not so easy on an Atlantic coast, where a sudden squall may drive the ship into danger, where the landmarks are not so conveniently spaced, and where visibility is seldom as clear as it is under blue Mediterranean skies. It was a fearful, nerve-wracking business, creeping around the shore of the Bay of Biscay.

It is possible that Pytheas cut straight across the Bay, for soon afterwards he did sail boldly out of sight of land when he crossed the English Channel at its widest part to reach the peninsula of Land's End. He may, too, have ventured out into the featureless ocean in the far north, where he encountered the sea ice. It is difficult for us to imagine how bold a venture that was. It was almost like stepping off the edge of the earth.

The reason why sailors preferred to stay in sight of land was that they had no other reliable way of knowing where they were or where they were going. They did not even have a compass, and that could lead to odd mistakes. An army once set out to conquer a people in the west, marched all through the night, and at dawn found they were marching directly towards the rising sun – the wrong direction completely. It is true that Pytheas had a way of measuring latitude (the distance north-south, between equator and pole). He had calculated the latitude of Marseilles with great accuracy, but his method involved making lengthy observations at the solstice, which only occurs twice a year (the longest and shortest day). It also required a large and cumbersome instrument like a sundial. This was not much use to an explorer and none at all on board a ship.

Navigators in Pytheas's time relied on 'dead reckoning' – really no more than intelligent guesswork. It might work well enough for an experienced sailor sailing from one headland to the next in familiar waters. It was rather unreliable in an unknown sea with no landmarks in sight. Pytheas's big mistake over the size of Britain is a good example of the unreliability of dead reckoning in strange waters. (However, some of Pytheas's measurements were very accurate; his estimate of the distance between the north of Scotland and the port of Marseilles was almost exactly right.) Dead reckoning was the only way of measuring longitude (distance in an east-west direction), but longitude could not be measured accurately by instruments until over two thousand years after Pytheas, and this problem concerned explorers up to the time of Captain Cook.

Another difficulty faced by all explorers, at least until the invention of refrigerators, was the storage of fresh food. Pytheas was less troubled by this than the explorers of the age of Columbus, because he was seldom out of sight of land for long and could get fresh water, fruit and vegetables more easily. He carried wine and olive oil in earthenware jars, which kept for a long time, and he probably carried both flour and grain in similar jars. He may even have taken seed to plant on the way, producing grain to reap later. For he would have had time to wait for the harvest during his long absence.

The diet of people in the ancient world was simpler than ours. Meat was a rare luxury, and Pytheas would probably not have thought of taking any even if he had some way of preserving it. The soldiers who conquered a great empire for Rome lived on a mainly vegetarian diet.

Altogether, Pytheas and his comrades were not so poorly equipped for a long voyage of exploration as we might think. If they had returned to life 1500 years later they would have found the changes surprisingly small. Ships were very little better. Methods of navigation were still few, though the compass had arrived, and the problems of fresh food and water were the same.

All that does not reduce the achievement of Pytheas. He was the first really scientific explorer, and he added enormously to the geography of his time. He is the one man of the ancient world who stands, as a great traveller, on a level with Columbus, Magellan and the other heroes of the age of discovery, nearly two thousand years after his time.

Eric the Red

The Vikings go West 960 AD

Vikings is the name given to the inhabitants of Scandinavia – Danes, Norsemen (Norwegians) and Swedes – in the early Middle Ages. The name means 'sea raiders', and for a period of about two hundred and fifty years (from the late eighth to the early eleventh century) Vikings raided the coasts of Europe.

Although the Vikings began as raiders, speeding across the sea in sleek warships to attack some peaceful village on the coast of Britain or Ireland or France, they soon became invaders and settlers. In Scandinavia there was not enough good farming land to support the growing numbers of people. So they spread abroad, at first seizing what they wanted by violence, later settling as peaceful farmers, and finally becoming natives of the country they had settled in.

The Vikings who sailed south-west and settled in England and France were mostly Danes. The Swedes moved eastward into Russia. The Norsemen sailed towards the west.

In the ninth century the Norsemen reached Iceland. In less than one hundred years a large and thriving community was established there.

From parts of western Iceland it is just possible, on a very clear day, to see in the distance the glinting ice-covered mountains of Greenland, about 300 kilometres away. Icelanders knew that Greenland existed long before they went there. Or perhaps they did visit it and found it unattractive. The Vikings did not go on their travels just because they were interested in discovering new countries. They needed more practical reasons. Eric the Red had one.

Outlawed from Norway

Although the Vikings were not all such blood-thirsty barbarians as their enemies said, violence was part of their everyday life. When a Viking spoke of 'justice' he meant what we should call 'revenge'. Family feuds, in which one murder followed another, were common.

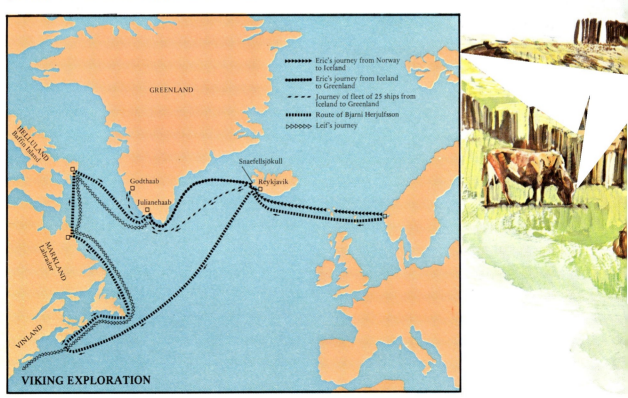

GREENLAND

HELLULAND
Baffin Island

Godthaab

Julianehaab

MARKLAND
Labrador

VINLAND

Snaefellsjökull

Reykjavik

▶▶▶▶▶▶▶ Eric's journey from Norway to Iceland
●●●●●●●● Eric's journey from Iceland to Greenland
- - - - Journey of fleet of 25 ships from Iceland to Greenland
▮▮▮▮▮▮▮ Route of Bjarni Herjulfsson
▷▷▷▷▷▷ Leif's journey

VIKING EXPLORATION

So it was nothing very unusual when, about 960, the father of Eric the Red was forced to leave Norway with his son because of some killings they had been involved in. They went to Iceland, which was by that time an established nation. Eric's father died in Iceland, and soon afterwards Eric got into a long and murderous argument like the one which had caused his father's banishment from Norway. The result was the same too. Eric was declared an outlaw and forced to leave Iceland.

Animals provided food and other necessities.

Eric's farm was built from stone and strips of turf. The house was about 7½ metres by 16½ metres, and had very thick walls. There were also several outhouses, four barns and two byres.

A bronze plaque, probably connected with the worship of Odin. Eric the Red was a pagan when he died, but Leif tried to convert the Greenlanders to Christianity.

He prepared his ship and, with a group of his supporters, he set out across the Denmark Strait. He reached the east coast of Greenland without difficulty, but found a hostile, ice-bound country. So he sailed south, and came to the place where Julianehaab is now. Here the land was more pleasant and the weather was no harsher than in Iceland.

During the next three years Eric explored much of the western coast and discovered the site of the future Godthaab. He returned to Iceland to tell of his discoveries and the following summer he sailed to Greenland again, this time to settle there for good. Many people went with him, but by bad luck the fleet was caught in a severe storm. Out of 25 ships that set out with Eric only 14 reached Greenland, although some others got back to Iceland safely.

Eric made his home at Eriksfjord (Julianehaab), at a farm called Brattahlid. Others settled nearby, while a smaller group went up the coast and founded the future town of Godthaab.

Greenland and Vinland

Eric called the country Greenland because he wanted people to come and live there. At that time the climate was milder than it is today, and even now a small part of Greenland really is 'green'. In the south and south-west, between the icy mountains of the land and the grim grey waves of the sea, is a coastal strip where grass grows and animals may graze.

The Greenlanders survived and prospered. Eric the Red, their leader, had a farm which contained four barns and had stalls for 40 cattle. His comrades netted fish from the many rivers and fjords. They hunted large animals like reindeer, seal and walrus. Although they had to import most of their corn to make bread, and their metals to make tools and weapons, they had goods of their own to sell in exchange: ivory from walrus tusks, which was used for knife handles and for carving ornaments like chessmen, skins and oil from walrus and whales, and strong ropes plaited from strips of walrus hide.

The Greenland colony lasted nearly five hundred years. But in the fourteenth century the climate quite suddenly turned colder. The Eskimos who had lived farther north were driven south. The severe climate and competition from the Eskimos, who were better able to survive in a cold land, caused the decline and disappearance of the descendants of Eric.

Soon after the first Greenland settlers had left Iceland, a young man named Bjarni Herjulfsson arrived from Norway to visit his father. But his father had gone with Eric to Greenland. Bjarni was rather put out, but he decided to follow.

Bjarni and his crew sailed for three days. Then fog closed in, the wind changed, and they were lost. When the sun came out again they could at least check their position by the sun, but they no longer had any idea where they were. Soon they sighted land, but Bjarni thought it could not be Greenland because it had no mountains and he could see thick woods. They sailed on, and came to another land, but this too was covered with trees, which do not grow in Greenland. They put out to sea again, sailing before a south-west wind, and came to an island. That certainly was not Greenland. Four more days brought them to a fourth unfamiliar coast, but this time they were in luck. Bjarni recognised the place from a description, and they landed quite close to the spot where Bjarni's father had his farm.

The eldest son of Eric the Red, whose name was Leif, bought Bjarni's boat, engaged a crew, and prepared to visit the new land which Bjarni had seen. A big man, tall and strong, and with unusually sharp eyesight, Leif was also clever, and a good deal less hot-tempered than most men of his race. He earned the nickname Leif the Lucky after he had rescued some people from a boat stuck on a reef. The name meant that he was lucky for others as well as fortunate in himself.

The first land that Leif and his comrades came to was like one great barren rock. They called it *Helluland* ('Slab-land'). It was probably Baffin Island. They sailed on to the next place, which looked more promising as it was wooded. They called it *Markland* ('Forest-land'). It was probably Labrador. After a quick look round they sailed on. They rounded a headland and went up a river to a lake. Here they made a longer stay. They put up temporary shelters – low, rough walls of stone and turf which could be roofed over temporarily, perhaps with an old sail from the ship. Later, they built more permanent houses, for they had decided to spend the winter there.

One of Leif's men went off by himself one day and returned with exciting news. Not far away he had found wild vines, with grapes growing on them. It was this that made Leif name the country *Vinland* – 'Wine-land'.

Unfortunately, we do not know where Vinland was. It may have been somewhere in New England.

Leif and his men set to work getting a cargo ready to sell in Greenland. They filled one small boat with grapes and timber – something in very short supply in Greenland. Their arrival in

Tyrkir, a German member of Leif's crew, found vines, with grapes, growing wild. Leif called the land 'Vinland' which means 'Wineland'.

Greenland caused great excitement. Thorvald, another son of Eric the Red, borrowed Leif's boat and next year visited Vinland himself. He and his thirty companions stayed in Leif's houses for the winter. There was no shortage of food thanks to good stocks of fish. One trick of the Greenlanders was to dig trenches on the beach at low tide. The tide came in, filling the trenches, and when it went out again, halibut were left trapped in the trenches.

In the spring Thorvald's men went exploring. They found a beautiful stretch of country, with woods growing almost down to the white sandy beaches. They also found signs of men.

In the grim world of the Vikings, strangers were likely to be enemies. Thorvald's men attacked the American Indians. Later there was a sea fight, but the Indians soon gave up. Only one of the Greenlanders was injured, and that was Thorvald himself. An Indian arrow had passed through the narrow gap between the line of shields along the ship and the ship's side. It struck Thorvald in the armpit and went deep. 'I said I would like to settle here,' said the courageous Thorvald, 'and it seems I shall have my wish.' He died and his friends buried him in the soil of Vinland.

A third son of Eric set out for Vinland the following year to bring back Thorvald's body, but he was blown off course. His ship reached the other Greenland colony, at Godthaab, but he died there of some epidemic that killed many of the colonists.

His wife, a beautiful woman named Gudrid, returned to Brattahlid where, since Eric's death, Leif the Lucky was head of the clan. She married again not long afterwards. Her second husband was Thorfinn Kalsefni, a rich man and as adventurous as any member of Eric the Red's family. It was Kalsefni who established the first permanent colony in Vinland. The colonists took cattle with them, including a bull so the animals might breed, and they traded milk to the Indians in exchange for furs. A stranded whale on the beach provided meat and fuel for the first winter.

Unfortunately the Greenlanders could not stay on good terms with the Indians, whom they foolishly despised. They were too few and too weak to survive in North America as long as the Indians remained hostile, and Kalsefni soon returned to Greenland. The Viking colony never took firm root, and when America was re-discovered by Columbus nearly five hundred years later, Vinland was long forgotten.

The Vikings' war-like nature led them to attack any strangers of whom they were suspicious. In one such encounter Thorvald was killed.

The sail on a Viking ship could be used to make a tent-like cover.

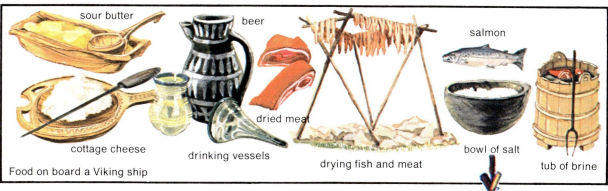

sour butter

beer

salmon

dried meat

cottage cheese

drinking vessels

drying fish and meat

bowl of salt

tub of brine

Food on board a Viking ship

Viking ships

The Vikings were able to travel so far because they had excellent ships and were first-rate seamen. Life on board was hard and uncomfortable, as life on small ships usually is. But the Vikings expected nothing better. There were no cabins, though sometimes a tent might be erected on a wooden frame. Men often slept in a kind of leather sleeping bag, two or three to a bag because it was warmer. Their food was dried fish and meat, sour butter and something like cottage cheese, with beer to drink. Most ships carried a big bronze pot for cooking, but it was probably used only when the crew were able to land and start a fire on the beach. Trading ships had one or two small boats, which were either carried on deck, where they might act as shelter in wet weather, or were towed behind.

The Vikings were the first European people to make regular long voyages across open sea with no land in sight. They had no compass and sailed by the sun. They knew that the height of the sun at midday is an indication of latitude, and they could calculate this with the aid of tables worked out by a scholarly Icelander in the late 10th century. They probably used the stars too, but it is not easy to see the stars on a light summer night in the far north. When neither sun nor stars were visible the Viking ship was, for the time being, lost.

Viking navigators steered by a bearing dial. This is a disc of wood with a stick through the centre acting as a handle. Around the side of the disc are notches; one of the notches represents south. At noon, the navigator lines up the south notch with a point on the horizon where an imaginary line drawn down from the sun would meet it. This lining-up was made easier by the spike on top of the central stick. The navigator set the course he wanted to follow with the pointer, then he steered the ship in line with the pointer.

At night he might be able to use the Pole Star

to give him north. He might also be able to take a bearing on the rising sun at dawn, for there were tables which told him at what point the sun rose at different seasons.

The ships of the Vikings, their skill in navigation, and the boldness of their ocean voyages were far ahead of the rest of Europe. There were no seamen to equal them until the Portuguese in the late fourteenth century, five centuries after Eric the Red first sailed to Greenland.

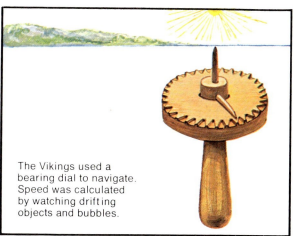

The Vikings used a bearing dial to navigate. Speed was calculated by watching drifting objects and bubbles.

Marco Polo

Twenty years in China 1271

Eight hundred years ago Venice was one of the most advanced states in Europe. Its prosperity came from trade, and its merchants were its richest and most adventurous citizens. The Polo brothers, Niccolo and Maffeo, were two typical Venetian merchants who, in the 1260s, set out on a long journey to the east in search of trade.

They went first to Constantinople (now Istanbul), which was then a Christian city open to trade with Venice. From Constantinople they decided to make a daring trip into Tatar territory. Everything went well until a war broke out in their rear and cut them off from the route home. As they could not go back, they went on, and eventually reached Bukhara in Persia. This town was a centre of trade for Russia, China, Persia and India, countries that were almost completely unknown to Europeans at that time.

Kublai Khan

The Polos stayed in Bukhara three years. Then they were invited to join a Persian diplomatic mission on its way to the Tatar emperor of China, Kublai Khan. This tremendous ruler was the greatest of the Tatar emperors since Genghis Khan, the conqueror who had led his people out of Mongolia to rule over most of Asia and part of eastern Europe.

Not only was Kublai immensely powerful, he was intelligent, curious about the world around him, and in many ways far more 'civilised' than a European ruler of that time. He was tolerant of other races and other religions, and although he ruled China, he did not destroy Chinese civilization, which was the oldest in the world, but he protected it and learned from it. Chinese culture remained undamaged under Tatar rule. Kublai was interested in all religions, and eager to see what miracles could be worked by the rival priests and holy men who appeared at his court. In this respect, Christianity was a little disappointing to Kublai Khan, yet he was glad to see the Polo brothers and treated them generously. He discussed European affairs with them and gave them a letter to take to the Pope when they set off for home.

When the brothers arrived back in Venice they found that Niccolo's son, Marco, who had been a child when they left, had grown into a lively young man of 17. They decided to take him with them when, fulfilling a promise to Kublai Khan, they returned to China.

The main reason why Marco is more famous than the two elder Polos is that he was the author of a book which told the story of their travels. (Marco called it, rather grandly, a *Description of the World*.) Marco himself was no writer, but he dictated his memoirs some time after his return to Europe, when he was a middle-aged man.

The journey east

The three Polos set out in 1271. They were not the first Europeans to travel in the Khan's empire, but we know almost nothing about other travellers and, as far as the rest of Europe was concerned, almost everything that Marco Polo reported was new. He was not an explorer entering undiscovered countries, but a merchant who happened to record a fascinating and almost unique journey.

The route of the Polos (as far as we know it) can be followed on the map. In Armenia they were deserted by the two friars whom the Pope

Crossing the Gobi Desert probably took the Polos about 30 days.

had sent with them as missionaries to China. They went on without them, to Tabriz near the Caspian Sea, and from there south to the port of Hormuz on the Persian Gulf, where it was so hot that some people spent part of the day sitting up to their necks in water. They had hoped to take a ship at Hormuz, but after examining the local boats, which had no nails or pegs but were held together by string, they decided to make the long trek by land across the whole breadth of Asia. This meant they first had to cross the high plateau of the Pamirs, then the deadly expanse of the Gobi Desert.

The high plain of the Pamirs is about 5,000 metres above sea level. Fortunately, the Polos crossed it in summer, and instead of bleak stretches of ice and snow they found good pasture.

Crossing the Gobi Desert was a more frightening experience, and three Italian merchants of the thirteenth century could never have managed this month-long journey without the

gold tablets given them as a passport by Kublai Khan, the help of the local people, and the existence of a regular track.

On their journey through the Tatar domains, the Polos met with many things strange to them though well-known to us. Few people believed Marco's account of a kind of coarse cloth made from a stone-like fibre which was proof against fire. But thirteenth-century Europeans had never come across asbestos. Marco was astonished to find that goods were bought and sold in the great Khan's dominions with pieces of paper. He thought this was as good a trick as the supposed ability of alchemists to turn base metal into gold. But paper money is familiar enough to Marco's successors. No less astonishing to Marco Polo were 'stones that burn like logs', in fact better than logs, for they could be kept burning all night. However, this would not have seemed strange to a merchant from London, rather than Venice. If the Polos had been to England, they would have known about coal.

Servant of the Great Khan

When the Polos finally arrived at Kublai Khan's summer palace in Shangtu, they received a warm welcome – in spite of the fact that they came alone. Marco, being young and intelligent, at once took the Khan's eye, and when Marco's father introduced him as 'my son and your servant', Kublai thought he meant exactly that. He took Marco into his service, where he remained for twenty years. On his side, Marco became a warm admirer of the Khan, who is the real hero of his book. He rose high in the Khan's service, even acting as a governor of a province for a time.

THE POLOS' JOURNEY TO THE EAST

Shang-tu

abriz

PAMIR MOUNTAINS

GOBI DESERT

Hormuz

MARCO POLO'S JOURNEYS FOR THE KHAN

▷▷▷▷▷▷ Marco's first journey for the Khan

▶▶▶▶▶ A later journey

Shang-tu
Khambalig (Peking)
TIBET
KARA-JANG
Kinsai (Hangchow)
Zaiton (Amoy)
Pagan
HAINAN
INDIA
Motupalli
ANDAMAN ISLANDS
MIEN (BURMA)
Maarbar
CEYLON
NICOBAR ISLANDS
BINTAN
BORNEO
LESSER JAVA (SUMATRA)
JAVA

Kublai valued the judgment of an intelligent and unbiased spectator like Marco, and for that reason he sent him, or allowed him to go, to report on provinces throughout his empire and beyond it. In the service of Kublai Khan, Marco Polo came to visit Burma, India, Ceylon, Indonesia – countries not seen again by Europeans for over two hundred years.

Marco's first journey for Kublai Khan took him south-west to Kara-jang, near Burma, and across the whole of China. On the way he saw many marvellous sights, like the bridge 300 metres long across the River Hung-ho. It had 24 arches and was wide enough for ten horsemen to ride abreast along it. For once Marco may have been speaking the exact truth when he said there was no other bridge like it in the world.

Marco Polo was presented to Kublai Khan at the summer palace in Shang-tu – a portable structure of bamboo held together with silk ropes.

Above: Marco was impressed by the Pulisanghin bridge over the River Hung-ho.

Below: Fires in the forests of giant bamboo in Tibet caused loud explosions and terrified the horses. They bolted in panic, and it was necessary to blindfold them and put iron shackles on their legs to keep them under control.

Below: Marco visited Kara-jang, a province ruled over by one of the Khan's sons, Essen Temur.

Marco reached Kara-jang at last, though he does not say whether he was still riding the same horse. The province had some unattractive inhabitants, especially what Marco calls 'huge serpents', probably crocodiles. The people, too, had some nasty habits, such as their use of poison and their custom of murdering house guests to bring luck to their house. Criminals often took poison to avoid the death penalty, says Marco, and to prevent that the authorities kept a bucket of dogs' dirt at the gaol. If a prisoner was suspected of taking poison the dog dirt was forced down his throat. Not surprisingly it made him sick. He brought up the poison, and so he was preserved for the executioner.

A later journey on behalf of Kublai Khan took Marco to Sumatra and Ceylon (now called Sri Lanka). In Ceylon he was fascinated by the trade in precious stones – no doubt he invested in a few himself – and he gave a detailed description of pearl divers at work among the oysters in the coastal waters of Ceylon. In Sumatra Marco found the people made a kind of flour from trees. When the thin bark and a layer of wood underneath were stripped off a pithy substance like flour was revealed. It was put into troughs of water and stirred with a stick, so that the pure flour sank to the bottom while impurities floated to the surface. The water was drained off, leaving the gooey flour mixture, from which cakes and other dishes were made. This was the first European contact with sago.

The finest of all the cities visited by Marco Polo was Hangchow. In his description of it, Marco shows both his clear observation of matters of detail and his tendency to exaggerate wildly. It is impossible to believe him when he says that the 'city of heaven' had 12,000 bridges. That is his way of saying a great many. On the other hand, his account of everyday life in this great city (which made Venice look like 'a dirty village') is down-to-earth and realistic. He liked the life here, which he thought very civilised with its boating trips and gracious carriage drives, but he was surprised by the custom of taking cold baths. The people told him it was good for the health, but he was relieved to find that many of the public bath houses also had rooms where hot water was supplied, for the benefit of soft foreigners who could not stand the cold-water treatment.

However much they enjoyed life in the Tatar empire, the Polos, especially Marco's father and uncle, who were growing old, must have been anxious to go home. For a long time Kublai Khan was reluctant to let them go, but in 1291 they were put in charge of a young princess who was to marry the khan of Persia. Marco Polo left China with mixed feelings, knowing he would probably never return.

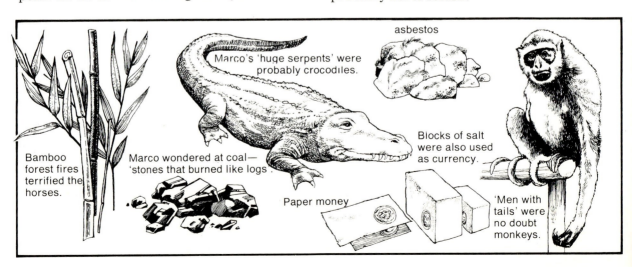

Bamboo forest fires terrified the horses.

Marco wondered at coal—'stones that burned like logs'.

Marco's 'huge serpents' were probably crocodiles.

asbestos

Blocks of salt were also used as currency.

Paper money

'Men with tails' were no doubt monkeys.

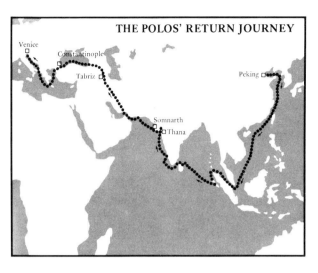

THE POLOS' RETURN JOURNEY

Venice
Constantinople
Tabriz
Peking
Somnarth
Thana

The journey home

This time they went by sea. Chinese ships were far better than the ships they had rejected at Hormuz on their outward journey. In fact the ship in which the Polos travelled was not much different from ships which were sailing in Chinese coastal waters less than fifty years ago.

They called at Vietnam, where Marco heard tales of men with long tails (monkeys, no doubt), and from there to Sumatra. As they were one of a fleet of fourteen vessels the pirates that preyed on shipping in the Indian Ocean left them alone. Piracy was a real family business in these parts, for the pirates took their wives and children on their cruises. However the pirates did not harm the crew of any ship they captured. They took the cargo and the ship but released the sailors, saying, 'Go, fetch another cargo. Then with luck we shall catch you again.'

At Hormuz the Polos completed the enormous circle of their travels which had begun twenty years before. They found the intended husband of the Tatar princess had died, but his son married her instead, and everyone seems to have been happy with that arrangement. The fleet returned to Cathay, and the Polos made their way overland to Italy. At first no one recognized them, for they had been presumed dead many years before, but when they ripped open the hems of their coats to release the precious stones hidden there, they were believed.

Unfortunately, though, Marco's stories – the true as well as the doubtful ones – were disbelieved. He was given the nickname 'Marco Millions' not because of the riches he had made in the Far East, but because of his tendency to exaggerate the size or number of everything he saw on his travels.

Columbus

Westwards to the Indies 1492

The age of discovery, when Europeans set out across the oceans of the world and found not only new countries but new continents began in the fifteenth century. First came the Portuguese, who sailed south into the Atlantic and eventually around South Africa to India. While they were laying the foundations of a commercial empire in the east, Spanish ships were carrying soldiers and settlers to the west – to New Spain, across the Atlantic Ocean. This Spanish empire, starting in the West Indies then expanding into Mexico, South America and Florida, was to make Spain the most powerful country in the sixteenth century. The man who began it, the most famous explorer in history, was Christopher Columbus.

The Spaniards knew him as Cristóbal Colón, but he was baptised Cristoforo Colombo, for he was by birth an Italian, from Genoa. When he was 25 he was shipwrecked off Portugal, and settled in that country. It was the best place for a navigator to learn his trade although, as a matter of fact, Columbus was a far from expert navigator when he set out on his first voyage of discovery. Some of the men under him were better navigators than he.

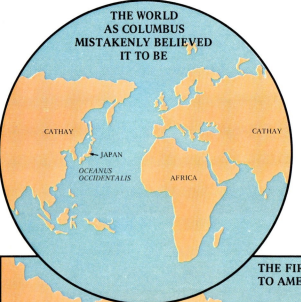

THE WORLD
AS COLUMBUS
MISTAKENLY BELIEVED
IT TO BE

CATHAY

CATHAY

→ JAPAN

OCEANUS OCCIDENTALIS

AFRICA

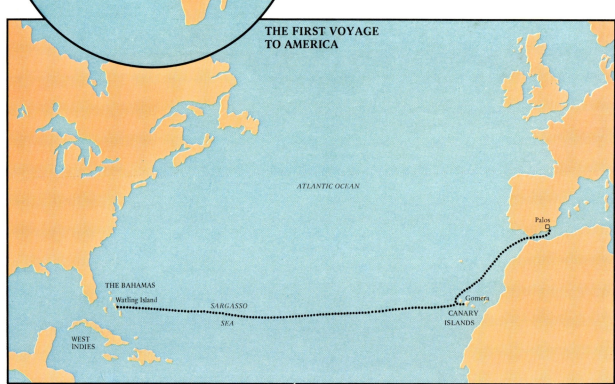

THE FIRST VOYAGE
TO AMERICA

ATLANTIC OCEAN

Palos

THE BAHAMAS

Watling Island

SARGASSO SEA

Gomera

CANARY ISLANDS

WEST INDIES

It was Queen Isabella who finally gave Columbus the backing for his expedition. Although she was not particularly attractive as a young woman *(see right)* she was one of the most intelligent and best educated women of her day. Like Columbus, she was deeply religious and there was a strong bond between them.

Ferdinand

Isabella

Search for the Spice Islands

Columbus was driven by ambition to seek greater fortune and adventure. Like many people of his time he was fascinated by the idea of the 'Indies', those spice-rich islands which the Portuguese were trying to reach via the Indian Ocean. Columbus had another route in mind. Instead of sailing east which first, as the Portuguese were discovering, meant sailing a very long way south, Columbus suggested sailing *west*. As the world was round the ship would eventually reach the Indies from the opposite direction. Columbus believed that the western route would be both easier and shorter.

He did not know that the double continent of the Americas bars the way from Europe to Asia. Strange as it seems now, the existence of America was unknown and unsuspected. Although Norsemen had settled for a time in Newfoundland, and other ancient sailors may have sighted the American coastline at one time or another, Columbus knew nothing of them. When he arrived among the islands of the Caribbean in 1492 he thought he had proved his theories correct and that he was in the (East) Indies. That is how the West Indies got their name.

Columbus always stuck to his belief that he was in the Far East. During his explorations in Cuba and Haiti he expected at any moment to stumble out of the jungle to find gilded Chinese palaces or Japanese pavilions.

Columbus also believed that the earth was smaller than it is, and that Asia extended much farther east than it does. These two mistakes combined made him think that the Far East was much nearer to Europe by the westward route that he proposed than it actually was. In fact, he estimated that the distance to Asia would be about the same as the distance across the Atlantic really is. (Columbus calculated that Japan lay 2,400 nautical miles due west of the Canary Islands.)

On board the *Santa María*

Before setting out on his voyage of discovery Columbus needed the backing of the state. The royal council of Portugal flatly rejected his plan, perhaps because Columbus, who did not suffer from the fault of modesty, demanded too big a salary, or perhaps because they thought his ideas were crazy. But he had better luck in Spain. With the approval of King Ferdinand and Queen Isabella – and royal cash – he could prepare in earnest for his voyage. It was a great

The *Santa Maria* was Columbus's flagship. She was shipwrecked on Christmas Day 1492.

relief to know that the way was clear at last. He had spent ten years of his life preparing for the day in August 1492, when his three ships sailed from the little port of Palos, on the Rio Tinto, on their voyage across the Atlantic – the Green Sea of Darkness, as Arab sailors called it.

The largest ship, the *Santa María*, had a crew of forty. The smaller vessels, *Niña* and *Pinta*, had twenty or twenty-five men each. They were mostly Spaniards and experienced sailors, but Columbus was an Italian – a foreigner – and a man who had never held a command before. This caused problems – grumbling, desertion and near-mutiny, for sailors were always inclined to be mutinous when far from home. But Columbus was well supported by most of his officers and, although he lost his largest vessel on a coral reef in the West Indies, his expedition had fewer troubles than many later ones.

Columbus had the explorer's powerful sense of curiosity. But what drove him more than that was ambition – a desire to be famous and rich, and to prove that the ideas he had been preaching with such conviction for the previous ten years were true. Some of his officers shared his feelings, but most of the company were sailing towards the setting sun in those crowded wooden ships because the government were paying them wages to do so. (Small wages, however.)

Other motives were at work among these men of the fifteenth century, including a kind of religious patriotism. In the very year that Columbus sailed, the last kingdom of the Moors (a Muslim people of North African origin) in

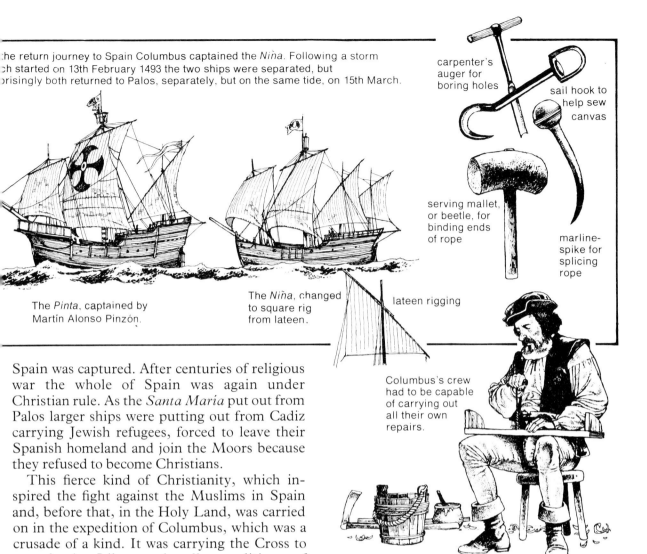

the return journey to Spain Columbus captained the *Niña*. Following a storm
:h started on 13th February 1493 the two ships were separated, but
orisingly both returned to Palos, separately, but on the same tide, on 15th March.

The *Pinta*, captained by
Martín Alonso Pinzón.

The *Niña*, changed
to square rig
from lateen.

carpenter's
auger for
boring holes

sail hook to
help sew
canvas

serving mallet,
or beetle, for
binding ends
of rope

marline-
spike for
splicing
rope

lateen rigging

Columbus's crew
had to be capable
of carrying out
all their own
repairs.

Spain was captured. After centuries of religious war the whole of Spain was again under Christian rule. As the *Santa María* put out from Palos larger ships were putting out from Cadiz carrying Jewish refugees, forced to leave their Spanish homeland and join the Moors because they refused to become Christians.

This fierce kind of Christianity, which inspired the fight against the Muslims in Spain and, before that, in the Holy Land, was carried on in the expedition of Columbus, which was a crusade of a kind. It was carrying the Cross to new lands. Like nearly all expeditions of discovery in that time, one of its chief purposes, as stated by its leaders, was to spread the Christian religion.

Taking advantage of the trade winds, Columbus's ships had a smooth, fast voyage, sometimes making nearly 200 nautical miles a day. They called at the Canary Islands to refill water casks and buy flour, cheese and pickled beef. Fresh food was not to be so great a problem on Atlantic crossings as on later, Pacific voyages. Columbus's ships also carried peas and beans, salted meat, sardines in brine, nuts, honey and dried grapes. They carried many other kinds of stores too, for ships had to carry everything they might need in the way of tools, weapons, ropes, spars and timber. There were already some repairs to make to the *Pinta*'s rudder, and Columbus took the opportunity to convert the *Niña*, his favourite, from lateen rig to square rig.

Columbus navigated mainly by dead reckoning. That meant he plotted his course according to his estimates of three factors: direction (indicated by compass), time (measured by sand glasses) and speed (which could not be measured, only guessed). At first he overestimated the speed of the *Santa María*, and his dead reckoning was therefore none too accurate.

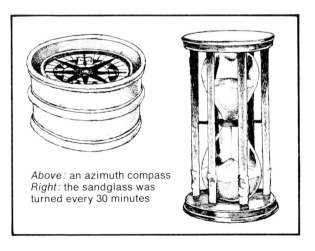

Above: an azimuth compass
Right: the sandglass was
turned every 30 minutes

Later, it improved. Columbus was not an outstanding navigator by the standards of his time. He was capable of making rather silly mistakes when observing the stars, and he did not know the new Portuguese method of calculating latitude by the altitude of the sun. What he did have was that precious gift, a navigational sixth sense, which enabled him, time after time, to arrive at exactly the place he intended after long voyages through uncharted waters.

Parrots from the West Indies caused a sensation in Europe.

One of the aims of Columbus's expedition was to bring Christianity to new lands, and there was a small religious ceremony on board the ships every morning. On the return journey to Spain, when the crew feared for their lives during a tremendous storm, the sailors vowed that if they were saved they would make a devout pilgrimage 'in procession in their shirts' to the first shrine to the Virgin Mary that they found.

Columbus had a poky little cabin in the *Santa María*, just big enough to lie down in, and his crew slept in any odd corner, on the open deck or anywhere they could find a space. They did not change their clothes during the voyage, nor take them off at night. Cooking was done, when it could be done at all, on a kind of portable open stove on the deck. Everyone had become rather tired of salted meat and dried beans by the time they returned to Spain.

The food on board Columbus's ships was simple, for not much cooking could be done.

wine · cheese · oil · flour · honey · dried fruit · bread · sardines · nuts

God was always present on the ships of those times. The day began at dawn with a kind of hymn sung by one of the ship's boys. (Every ship carried one or two boys among the crew, aged about 12 or even younger.) A boy also sang at the evening service, which was attended by all on board, and it was his job, too, to turn the sand glasses which measured time. It took thirty minutes for each glass to empty, and even this simple duty had a little religious chant to go with it:

The watch is called,
The glass floweth,
We shall make a good voyage
If God willeth.

A manatee. Surprisingly, these creatures were often mistaken for mermaids.

In spite of grumbles the voyage went well. We should have found the conditions very uncomfortable, but fifteenth-century sailors were used to nothing better. We might also have felt rather bored after a month at sea with little happening except the daily routine of the changing watches. But the men of Columbus's time were not so easily bored as we are. One of the most remarkable things about this famous voyage, in fact, was that it was so uneventful.

It was still the custom for European ships to sail close to the coast and not to venture far from land. Columbus had, by necessity, abandoned that custom by steering out into the middle of the ocean. For thirty-two days no land was sighted. This seemed strange and frightening, and the sight of mermaids (probably marine animals called manatees, or sea-cows, which do have a human look) was not much comfort. Columbus himself was growing worried because, according to his estimates, he should have sighted Japan. The crew were on the edge of open mutiny.

Just in time, signs of land appeared: floating branches with fresh leaves and flowers, birds flying overhead. The prophets of doom were silenced. In the early hours of October 12, by the ghostly light of the moon, the look-out of the *Pinta* saw a dim shape ahead. *'Tierra! Tierra!'* he shouted. Land!

It was Watling Island, a tiny place less than 16 kilometres by eight, in the Bahamas. At noon the ships anchored in a bay. A boat was lowered from the *Santa María*, and a few minutes later the Captain-General, Christopher Columbus, stepped on to the white coral beach. It was the end of the voyage and the beginning of a long story.

The return voyage

The return voyage was a far greater test of Columbus's ability as a navigator and as a leader of men than the outward voyage. He had only the two caravels – the *Santa María* having been wrecked. The winds were no longer so favourable, and at first the ships had to battle against head winds. They could not sail as near the wind as a modern sailing boat, and the *Pinta* was slowed down by a damaged mizzen mast.

The two caravels had some luck crossing the Sargasso Sea, where later ships were often becalmed. A full moon shone down on the gently heaving greenness of that weedy portion of the Atlantic. The caravels cruised through it with a gentle swish, helped by a following breeze, while far below, eels gathered for their mating rites.

Instead of trying to sail straight back to Spain, which would have meant contrary winds all the way, Columbus took a northward swing which, though of course he did not know it, is actually the best course. In the latitude of Bermuda the two caravels entered a zone of strong westerlies, which spurred them rapidly homewards. At one time they were averaging eleven knots, a very good speed for a modern sailing ship of that size.

As they approached the Azores the wind fell and progress dropped to a crawl. The ships drew near enough for Columbus to hold a discussion with the captain of the *Pinta* on what to do next.

Everyone except the Captain-General thought they were much farther south than they really were. Columbus, with his instinct for navigation, was right.

But they barely made it, for the weather suddenly turned violent. A force 9 gale almost overturned the *Niña*, as, with hardly a scrap of canvas showing, she struggled against towering waves that rushed down upon her as if intent on her destruction. The conditions were almost those of a hurricane, with violent counter-winds and cross-seas turning the ocean into a churning witches' cauldron of a place. Columbus took over the job of officer of the deck, watching each wave and warning the helmsman below to change course to meet it. One mistake, and the *Niña* would have been swamped, then someone else would have had to discover America.

The two vessels became separated in the gale, and they reached Spain independently, battered but safe. During the storm each man had promised God that he would go on pilgrimages to religious shrines if he got back safely, and even Columbus admitted later that he had been thoroughly frightened.

Columbus was to make three more voyages to the New World he had discovered. His later career was not entirely successful, and he had troubles ahead of him as formidable as the Atlantic storm. No one yet knew what would come of his discoveries, and no one knew that he had discovered an entirely new continent. But meanwhile, the discoverer received a jubilant, royal welcome. It was a time for thanksgiving and rejoicing, a time for dreams of a rich and glorious future.

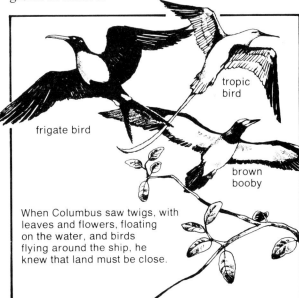

tropic bird

frigate bird

brown booby

When Columbus saw twigs, with leaves and flowers, floating on the water, and birds flying around the ship, he knew that land must be close.

31

Magellan

Journey around the World 1519

The greatest age of exploration was the period from the late fifteenth to the early sixteenth century. In that period Bartholomeu Diaz discovered the Cape of Good Hope, Vasco da Gama sailed to India, and Columbus crossed the Atlantic to discover the Americas. But the greatest voyage of all was the voyage of the

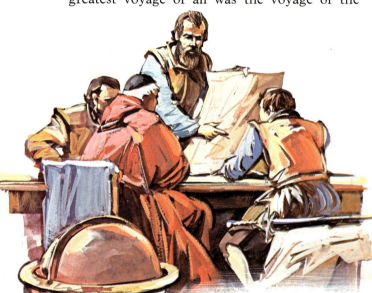

Spanish ship *Victoria* in 1519–22. The *Victoria* was the first ship to sail the whole way around the world, proving to any people who still doubted it that the Earth was round, not flat.

The expedition was not planned as a round-the-world voyage. The aim was simply to reach the fabulous Spice Islands of the Pacific by a westward route via South America. (The Portuguese had already made contact with the Spice Islands by the eastern route via India.) Four men were involved in the plan. The first was a rich banker and merchant, Cristóbal de Haro, who was born in Spain, lived in Portugal, and worked as agent for German merchant-bankers. In 1516 he moved to Spain, and there linked up with the second of the four, the Bishop of Burgos. A powerful politician as well as a priest, the Bishop had helped to organize Columbus's voyage in 1492 and had remained interested in the search for new ocean routes ever since. The third man was a scholar, Rui Faleiro, who was supposed to be an expert in astronomy and mathematics. He provided geographical knowledge, while Haro provided money and the Bishop provided the influence at court needed to gain royal approval for the expedition.

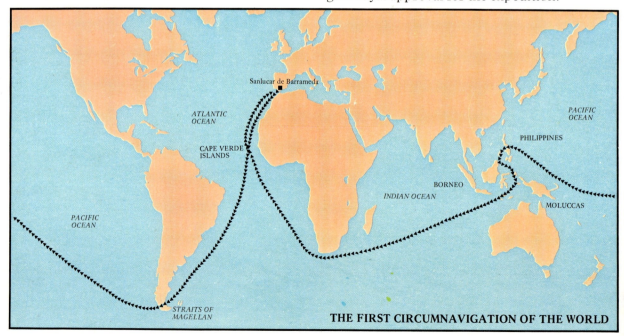

THE FIRST CIRCUMNAVIGATION OF THE WORLD

Santiago:

wrecked November 1519

Victoria:

round the world

Concepción:

scuttled in Philippines

Trinidad:

captured by Portuguese

San Antonio:

turned back during Pacific crossing

The *Victoria* was the second smallest ship in Magellan's fleet, but was the only one that completed the voyage round the world. At the start of the voyage she was captained by Luis de Mendoza.

Opposite page: The first voyage round the world was planned by four men: Cristóbal de Haro, the Bishop of Burgos, Rui Faleiro and Ferdinand Magellan.

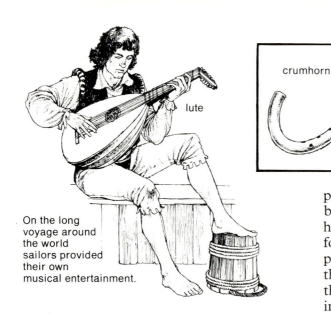

On the long voyage around the world sailors provided their own musical entertainment.

crumhorn   *tambourine*   *small drums*   *lute*

The man of action

The fourth man had to be a man of action, an experienced captain fit to command a large expedition, and preferably a man who had experience of the east, where the voyage was aimed. In 1517 the ideal man appeared.

Ferdinand Magellan was a Portuguese, the son of a squire. He had entered royal service and gone to India in a great expedition to set up Portuguese power there in 1505, when he was about 25 years old. The next ten years of his life were adventurous and dangerous and, for an ambitious man, disappointing. When he returned to Portugal and found he was to receive none of the rewards he expected for his services in the Far East, he felt insulted. More trouble followed when he was accused of stealing some supplies he had charge of during a campaign in Morocco. Without permission he returned to Portugal to put his case before the King. The King refused to listen and, although Magellan was later found not guilty of the charges, he applied for permission to take service under some other monarch. The King of Portugal insultingly told him he did not care what he did, so Magellan went to Spain.

He had already formed a plan for a voyage to the Spice Islands. Faleiro, whom he had met in Lisbon, had convinced him that a passage existed to the south of South America, which would provide a new route to the Spice Islands. From the reports of Francisco Serrão, a friend of his who lived in the Moluccas, Magellan believed that at least some of the Spice Islands were in the Spanish half of the world, according to the Treaty of Tordesillas (1494), which cheekily divided the undiscovered world equally between Spain and Portugal. In 1517 Magellan put his plan to the young man who had just become King of Spain, the young man known to history as the Emperor Charles V. Charles, eager for great deeds, welcomed it. In spite of the protests of the Portuguese, who did not accept that any part of the Spice Islands might fall in the Spanish part of the world, Magellan set sail in 1519.

Ships and supplies

Magellan had five ships, all very small by our standards. The *Victoria* looked something like Columbus's *Santa Maria*, and the other four were probably of the same type. According to the Portuguese ambassador, who went to look at them, they were not only small but old and unfit for a Mediterranean voyage, never mind a voyage across two oceans. However, the ambassador was probably exaggerating. He hoped the ships would sink anyway.

More likely, the ships were good sturdy merchantmen, though not newly built. There was nothing special about them, nor about other arrangements for the voyage. The instruments for navigation were no different from those carried by any ship of the time on a long voyage. Again like other ships, Magellan's vessels were equipped with tools such as saws, hammers and chisels so that they could make running repairs if damaged on some uninhabited shore. There were other, stranger objects on board, including a surprisingly large number of musical instruments. Fishing tackle too was provided, but this was not just for amusement. Without fresh fish everyone might have died during the long ocean crossing.

cod

Fish was the food most easily available for Magellan.

Discipline on board ship had to be very strict to combat mutiny such as occurred among Magellan's crew. Punishments that were then in use on board ships of all nations throughout the world would today be considered brutal.

flogging

keel-hauling

bilboes

cat o' nine tails

The gravest weakness, it soon turned out, had nothing to do with ships or equipment, but was a human problem. Like Columbus, Magellan was a foreigner in command of a Spanish expedition. There was no creature more proud than a Spanish gentleman, and the captains serving under Magellan did not take his orders easily. Magellan himself was a prickly character, and he refused to discuss his plans with his captains,

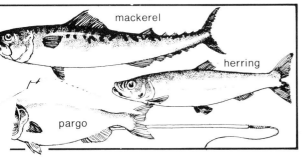

mackerel

herring

pargo

although it was the normal practice then to reach decisions by agreement among all the officers.

The crisis came to a head at a cold grim place in South America, where the first winter was spent. Mutiny broke out among officers and crew. Magellan, warned in time, kept control and defeated the mutineers. The chief Spanish officer was marooned on that deserted coast and other leaders were executed.

Magellan's Strait
Magellan was certain that a strait existed in the far south which would take his ship from the Atlantic to the Pacific. Why he should have been so certain is a mystery. No one had been there before and, so far as we know, Magellan had not even seen a map which showed such a strait, although Rui Faleiro believed the strait existed.

When Magellan discovered the strait that Rui Faleiro was so sure existed, he had no idea what navigational difficulties he had to overcome. Later sailors were to confirm that the passage from east to west is particularly perilous. The banks at the entrance to the strait are grassy and pleasant, but later the channel is narrow, with mountains on either side. The land to the south of the strait was inhabited by Onas Indians whose fires Magellan used to see at night. For this reason he called the place Tierra del Fuego – Land of Fire.

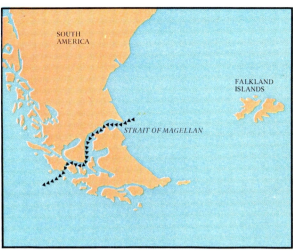

Anyway, Faleiro was right. Magellan found the strait, though it is not at all easy to find. The entrance is small, and easily overlooked. Once past the entrance, there is a fairly easy passage with low, grassy country on either side – pleasant country, Magellan thought. Then the strait appears to be blocked, and Magellan prepared to return, thinking that this was just another land-locked inlet like several he had inspected before. But he sent boats ahead to make sure, and they discovered that around a sharp bend the strait continued.

The second part of the strait is very different sailing. Ice-topped mountains loom on either side. The channel is narrow, the bottom rocky, so that anchors will not hold, and the prevailing wind blows like a fury from the *west*. Magellan was heading straight into it. At times he had to tow his ships with rowing boats. He had only three ships left now – one had been wrecked and another had returned to Spain. He was short of food and water, and his crew were miserable. They wanted to turn back, but Magellan pressed on. After 38 days – a shorter passage than many later ships – they limped past a gloomy-looking headland and suddenly found before them the open sea.

Crossing the Pacific

Because longitude could not be measured satisfactorily in the sixteenth century it was difficult to calculate east-west distances. Columbus had believed that the world was very much smaller than it really is, and Magellan believed the same thing. Therefore when he emerged into the Pacific Ocean he expected a voyage of a week or two – a month at most – before reaching the Spice Islands which, he knew, lay on the other side.

We know that the Pacific Ocean takes up one-third of the area of the Earth and is as large as all the other oceans put together. Magellan set out to cross it as though he were crossing the Mediterranean. The result of this mistake, which was made by all the maps of the time, was grim. For almost four months the ships were at sea without any kind of fresh food. The Spanish sailors ate ship's biscuit which had long before turned into an unrecogniseable dusty powder,

swarming with worms and smelling of the urine of rats. They drank stinking water that was dark yellow in colour. They stripped the leather coverings off the yards, soaked them in the sea for four or five days, toasted them over a fire, and ate them, chewing fiercely. They ate sawdust from the ship's planks. They scrambled in the dark holds to catch rats, which sold on board the *Victoria* at half a ducat each. (How they wished there were more rats to catch.)

Scurvy soon appeared. Men's gums swelled up, and they could not eat what 'food' there was. Nearly all fell sick eventually, and many died. The one piece of good luck they had was fine weather, which made them think that this ocean was well named 'pacific'.

At last they reached the island of Guam, and the long Pacific voyage was over. An Italian gentleman who sailed with Magellan and wrote an account of the voyage said that he did not believe any ship would again try to make the Pacific crossing.

After visiting the Ladrones – 'Thieves'' – Islands where the people could not keep their hands off anything they could carry away, the Spaniards sailed on to the Philippines. There Magellan, trying to establish Spain as a power in this new part of the world, foolishly involved himself in local politics. He told the King of Cebu that he would help him in a war against his neighbour, but the Spaniards were defeated and Magellan was killed.

Towards the end of 1551 the expedition finally reached the Spice Islands where the *Victoria*, now under the command of a Spanish officer, Sebastian del Cano, loaded up with cloves. Of the other two ships, one decided to return home the way they had come and the other was burned because there were not enough men left alive to provide her with a crew. Del Cano, a first-rate seaman, finally brought the *Victoria* home, crossing the Indian Ocean, rounding the Cape of Good Hope, and reaching Seville in September 1522. Out of 260 men who had set out with Magellan three years before, just 18 disembarked from the *Victoria* on that September day.

Navigational instruments

Navigators in the great age of European discovery – the age of Columbus and Magellan – had one very valuable aid, the compass. With it they could follow a course no matter whether the sky was clear or overcast. But the compass, like other instruments, did not work perfectly in practice. Compasses are liable to error caused by magnetic variation, a characteristic of the earth which was not understood in the sixteenth century. They could be thrown out if there were iron nearby and, as they were made by hand and

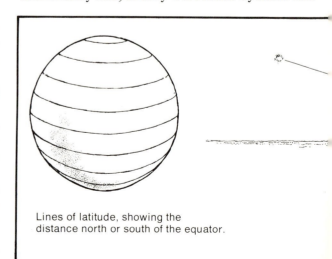

Lines of latitude, showing the distance north or south of the equator.

by a variety of craftsmen, they were sometimes rather rough and unreliable.

Following a course is only half the navigator's job. The first necessity is to know where you are: it would have been no use Magellan discovering his strait if he had been unable to tell others where it was. Explorers have to be able to tell their latitude and longitude.

To find out longitude exactly it is necessary to have an accurate clock or chronometer. This was not invented until the eighteenth century. There were accurate clocks before that, but they were pendulum clocks and no good at sea. Measuring longitude therefore remained a problem until the days of Captain Cook. Although some very accurate results were obtained by dead reckoning from the time and place of departure (by estimating speed and direction), on a long voyage errors tended to build up.

Measuring latitude was easier. Latitude is calculated by measuring the altitude – the angle above the horizon – of the sun or a fixed star. The Pole Star, which also indicates north, was useful for this. Its altitude grows steadily less as a ship sails south. In the Southern Hemisphere it is no longer visible, and navigators had to use another star.

The navigator in Magellan's day could use one of three instruments to measure latitude. The simplest was the quadrant, the ancestor of the modern sextant. The quadrant was a metal plate the shape of a quarter-circle. Around the curved edge a scale of degrees, from 0 to 90, was marked. Along one of the straight edges were two pinhole sights, and from the point hung a plumb line (a piece of string with a weight attached). To find the altitude of the Pole Star, the navigator lined up the star in the sights; the plumb line

hanging from the point indicated on the scale along the curved edge the number of degrees above the horizon. This reading gave the latitude, after a small correction to allow for the fact that the position of the Pole Star changes somewhat.

The quadrant could not be used to sight the sun because it is almost impossible to look directly at the sun through a sight. An astrolabe was used instead. This was a brass disc with degrees marked around the edge. It could be hung from a ring on its edge. A metal arm with a sight at each end was pivoted in the centre of the disc. To take a sight of the sun, the navigator hung the astrolabe in a convenient place and turned the arm until the light of the sun shining through the sight nearest to it fell exactly on to the sight at the other end of the arm. The correct degree of altitude was then indicated by the arm on the scale around the edge of the instrument.

For the seaman, neither the quadrant nor the astrolabe was easy to use for the simple reason that, even in the calmest sea, a ship is never completely still. Really accurate readings had to be taken on land.

The third instrument used by navigators to measure latitude, the cross-staff, was better designed for use on board ship because it did not depend on being perfectly vertical like a plumb line. The longer arm, which was marked with a scale, was lined up with the horizon, and the shorter arm moved along it until its top was lined up with the Pole Star. Its position along the scale indicated altitude.

The cross-staff could only be used to take a sight of a star. A later development, the back-staff, allowed a sight to be taken of the sun, but it was not invented until after Magellan's time.

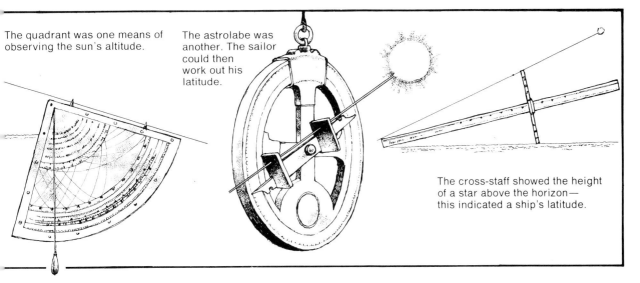

The quadrant was one means of observing the sun's altitude.

The astrolabe was another. The sailor could then work out his latitude.

The cross-staff showed the height of a star above the horizon— this indicated a ship's latitude.

La Salle

The great river 'Mississippi' 1681

Robert Cavellier, Sieur de la Salle, was the son of a wealthy family from Normandy. He was educated to become a Jesuit but, being the type of man who cannot bear to take orders from anyone, he found the rules of the Jesuits too strict for him. Unfortunately, under the law of France at that time Jesuits forfeited all their personal wealth, and at the age of 21 La Salle found himself penniless. He sailed to Canada in 1666 with the hope of replacing the fortune he had lost. The French at this time saw the New World as a place for trading, and many energetic and hopeful Frenchmen had gone there to seek their fortune.

Iroquois

Sioux

Like so many of La Salle's other projects the *Griffin* was doomed to failure. Although she collected a full cargo, she sank before she could return to Niagara Falls.

THE GREAT LAKES OF NORTH AMERICA

LAKE SUPERIOR

LAKE MICHIGAN

LAKE HURON

LAKE ONTARIO

LAKE ERIE

A man of vision

La Salle arrived in Canada at the age of 23, seven years before Joliet, another Frenchman, led an expedition down the Mississippi, as far as its junction with the Arkansas River. La Salle became a fur trader and a successful one. But he was never content with humdrum ordinary life nor with modest success. He was always dreaming of bigger and better things. He travelled widely in Canada and planned to follow Joliet's route, and carry on to find the mouth of the 'great river' – the Mississippi. He knew the importance that the French government attached to this waterway, and the governor of New France, the Count de Frontenac, was actively encouraging exploration, hoping to gain control of what might be a new route either to the Far East or, at least, to the sea.

Fur trappers traded their pelts for whatever they needed.

shoes

whisky

WHISKY

pistol

SALT

GRAIN

salt

grain

tobacco

During his early travels La Salle discovered the Ohio river and followed it into what is now Kentucky. He ranged all over the Great Lakes region and, after a visit to Paris, he received from the French government some land where Kingston, Ontario, now stands. This lay on the trade route for furs coming towards Montreal from the west, and La Salle was able to take his pick of the furs at a low price.

It was at Fort Frontenac, named by La Salle in honour of the governor of New France, that he conceived the idea of building a sailing ship which would ply the Great Lakes, buying furs from the Indians, and deliver her cargo to a depot at the eastern end of Lake Erie. Such a method would vastly increase the quantity of the Great Lakes fur trade, which was currently shipped entirely in Indian canoes, and would cut down the time taken for delivery to the market in the east. Most important, the money that La Salle would make from the venture would subsidise an expedition to the south. But first, a ship would have to be built and a permanent trading post established in Iroquois territory, near Niagara Falls.

The Niagara journey

In 1678 La Salle travelled to Niagara. The existence of the Falls was already well known, but the area had not been explored because of the hostility of the Iroquois. With him La Salle took workmen, including carpenters recently arrived from France, a missionary named Father Hennepin, and La Salle's best friend, Henry de Tonti. Tonti was known as 'Iron Hand' because he wore a metal hand to replace his own which had been blown off in a battle against the Spaniards some years before.

While the *Griffin*, La Salle's ship, was being built at Niagara, he returned to Fort Frontenac for more supplies. Difficulties were building up – lack of money, rebellious workmen, dangerous Indians and, worst of all, the attacks of his many enemies in Montreal who, out of jealousy, wanted his plans to fail. La Salle, whose health was as weak as his courage was strong, struggled on against his numerous difficulties, and the *Griffin* was finally launched successfully in Lake Erie. A flat-bottomed, 50-ton vessel like a squat sailing barge, to a European eye she was an ugly craft, but the Indians admired her and called her the 'winged canoe'.

Through Lakes Erie, Huron and Michigan the *Griffin* sailed, stopping often to trade with the Indians. Her holds soon contained enough beaver pelts to make hats for every gentleman in France. La Salle, who was always in a hurry, sent her back to Niagara while he, Tonti and a few others prepared for the journey down the Mississippi. With canoes, they planned to travel down Lake Michigan to the site of the modern city of Chicago, where they would build a base.

Autumn was beginning, and the lake was stormy. Several times the canoes had to take shelter on the shore from surf that would have pounded them to sawdust. One capsized, and the occupants were narrowly saved from drowning. Cold, wet and hungry, they struggled on. Nika, a Shawnee Indian who had become La Salle's loyal attendant, could not find any game, and the men had to eat what was left on the bones of an animal the buzzards had pecked over. At last they reached their destination, with La Salle more silent and gloomy than usual because of a rumour that the *Griffin* had sunk on her way back to Niagara with her cargo of furs.

He decided to push on south in hope of finding friendly Illinois Indians on the St Joseph River. Winter had gripped the land, and conditions for travelling were cruel. Hands became frozen to the handles of canoe paddles, moccasins had to be tucked under the body at night or they were frozen hard as iron by morning. One night La Salle got lost in the frosty woods. He found the camp of a lone Indian hunter, where he spent the night while the camp's owner, not daring to approach, paced about in the trees, amazed at the boldness of his unwelcome guest. When daylight came, La Salle found his way back to camp.

Food was still the biggest problem, in spite of Nika's skill as a hunter. One day – a great stroke of luck – they found a bison stuck in a bog, shot it, hauled the carcass out, and were soon broiling steaks. Otherwise, they had to make do with the odd stringy goose or swan until they came up with the Illinois.

Unfortunately, the Indians were not quite as friendly as La Salle had hoped. La Salle's enemies in Montreal had managed to get messages even this far to make his task more difficult. The base he built in the Illinois country he named, with some bitterness, Fort Crèvecoeur – Heartbreak.

Here he planned to build a second *Griffin* which would sail on the Mississippi. Racing against time as always, he split his party into three. Tonti was left in charge at Fort Crèvecoeur. A second party was sent to explore the upper Mississippi, while La Salle himself went back to Fort Frontenac, checking on the

Nika and La Salle existed mainly on 'Indian corn'

They had to be dragged on sledges, unless a passage could be broken for them with sticks. Eventually they gave up the canoes altogether, and relied only on their feet. In places where there was no dry ground at all they slept, somehow, in the crooks of trees. Sometimes they built shelters with birchbark peeled from the trees and draped over a framework of sticks.

Their main food was, as always, maize, or 'Indian corn' (sweet corn), which was roughly ground between two stones and boiled. They seasoned it with meat or fish if they had any. All of La Salle's companions became ill on the journey. Even the tough Nika was spitting blood. But La Salle himself, an ordinary-looking, rather slight figure of a man, kept them going. After 65 days and 1,600 kilometres they reached what seemed like the luxurious comfort of the log-built camp at Niagara. There La Salle learned that the *Griffin* had definitely been lost, that a storeship containing goods ordered from France had also been wrecked, and that his enemies had stopped a group of his workmen travelling to Fort Frontenac. Meanwhile, at Fort Crèvecoeur, Tonti's men mutinied, destroying the base and the half-built ship, and plundered other trading posts established by La Salle. No wonder he felt Fate was fighting against him.

The journey down the Mississippi

By sheer determination La Salle fought his way out of the net that surrounded him. He talked his creditors into prolonging their loans and borrowed more money from his family. He pursued the vicious mutineers and captured or killed all of them. He re-established friendly relations with the Indians, whom he regarded more highly than did most Frenchmen of his time. After he had almost given Tonti up for dead, he found him again, and together the two friends went to Montreal to put before Governor Frontenac their scheme for a French and Indian expedition down the Mississippi. Frontenac shared La Salle's vision, and gave him his support.

Late in 1681 the expedition – 23 Frenchmen and 18 Indians (plus some of their families) – set off from Fort Miami, the camp at the south of Lake Michigan. They walked untidily in single file along the frozen Illinois River. Their canoes were carried on sledges, along with the usual equipment of guns and cooking pots, blankets and trade goods, axes and knives. Before they reached the Mississippi itself they were able to launch the canoes, and they made rapid progress.

fate of the first *Griffin* on the way. This journey, about 1,600 kilometres during the difficult conditions of the spring thaw, was the hardest of all the many journeys made by La Salle in North America.

The way passed through forest and marsh and over stony mountains, often in pouring sleet. The temperature dropped sharply at night, and the ground turned into a slimy, icy morass. The trails were unknown, the local people unfriendly. Often they had to sleep in the open, without food, keeping constant watch. During daylight they marched, carrying on their backs their blankets, spare clothes, cooking pots, axes, guns and gunpowder, and leather to make new moccasins when the old ones wore out.

At times they were up to their waists in near-freezing water, but the streams were often too choked with ice for the canoes to be paddled.

The journey was not hard. The Indians on the banks were friendly, and as the cold, dark forests gave way to warm, open plains, game was easily shot. The river itself provided fish. Soon they were past the point reached by Joliet eight years before, and still everything went smoothly. La Salle perhaps began to wonder if Fate had relented. It seemed that he might, at last, discover the emptying point of this 'great river'.

By the end of March they were in the warm, almost tropical south, where mossy growths hung lazily from the trees and noisy insects whirred and croaked in the thick green undergrowth. In the customary manner of European explorers La Salle claimed every village where they stopped for the King of France, while the inhabitants stood around cheering approvingly, not having the slightest idea what was going on. At one place the French were frightened by what

Chicago □
□ Joliet

Illinois River

St Louis □

Mississippi River

New Orleans □

THE MISSISSIPPI RIVER

The journey down the Mississippi (right) to the Gulf of Mexico was pleasant and easy in comparison with the rigorous conditions on the way to Fort Frontenac from Fort Crèvecoeur (below).

looked like a war party, but when La Salle held up a peace pipe all became friendly.

The canoes drifted on, passing alligators which, they were astonished to discover, were born from eggs, like chickens. The current slowed, and they found themselves in the sultry, marshy country of the Mississippi delta. The river divided into three, and La Salle divided his party likewise, each taking one of the channels. On 9 April, 1682 they saw before them the broad, open waters of the Gulf of Mexico.

The death of La Salle

La Salle returned to France with the news that he had claimed a vast new territory named Louisiana in honour of King Louis. He was warmly received. The French government decided to establish a colony at the mouth of the Mississippi, to reinforce the French claim to the river and the land through which it passed, and La Salle was the natural choice to take charge of it. In 1684 he sailed from France with a fleet of four ships on his last adventure.

Almost everything went wrong. La Salle quarrelled with the naval commander of the fleet, who later deserted him, and became very ill in the West Indies. His illness affected his mental state as well as his physical strength, and he never regained his old determination and strength of character. When land was sighted in the Gulf of Mexico the ships turned towards the west to look for the mouth of the Mississippi, although in reality they had already passed it, and it lay to the east of them.

They never did find it. With part of the expedition, La Salle landed in Matagorda Bay where he attempted to found his colony. But the storeship was wrecked, and most of the vital supplies, including building tools, were lost. The little colony, made up largely of misfits and near-criminals, was stuck on the coast of Texas between the broad expanse of the Gulf on one side and the vast near-desert on the other. Discipline began to break down. Many died. Petty quarrels broke out among the survivors. Nika, the Shawnee, was savagely murdered as he slept, by the man who was supposed to be the expedition's doctor. La Salle himself was ambushed by two of his men, and shot dead without warning.

One or two of his men reached the Mississippi and struggled upstream to the camp where Tonti was waiting, in vain, for the arrival of his old friend. The wretched colony of Matagorda soon disappeared without trace.

Cook

The Southern Continent 1768

In 1768 the British Admiralty sent Lieutenant James Cook on a voyage of exploration. It was to be one of the greatest voyages ever undertaken, thanks largely to Cook's genius as a commander. The voyage had several objectives. The first was to observe a rare astronomical happening – the planet Venus passing across the face of the sun. Scientists had calculated that the best place to make the observation was the newly discovered Pacific island of Tahiti.

More important to the Admiralty was the second purpose of the voyage – the search for *Terra Australis* – the southern continent.

For hundreds of years the legend of a great continent occupying a large part of the southern hemisphere had persisted in the minds – and maps – of geographers. Looking at the globe, men thought there was bound to be a huge land mass in the south to balance the mass of Europe and Asia in the north, and this land mass was believed to extend almost as far as the equator. In spite of many pioneer voyages in southern waters, this idea was still flourishing as late as 1768. Cook's task was to discover if there was any truth in the legend. A quiet, common-sense Yorkshireman, Cook himself never said directly what his own ideas about the southern continent were, but he probably doubted its existence.

Portuguese and Spanish, and more recently French and English captains had sailed through many parts of the Pacific, but much remained to be discovered. 'New Holland', better known to us as Australia, had been discovered a long time before by the Dutch. Its western coasts had been surveyed, but its eastern part was unknown. No one knew how far it extended. The Dutch captain Tasman (for whom Tasmania is named) had also sighted New Zealand, and had sailed a short distance along its north-west coast. The supporters of the great southern continent believed that this land would prove to be an outlying cape of *Terra Australis*.

The captain and the ship

In the eighteenth century, birth and family relationships counted more than talent in most walks of life, but Captain Cook was the son of a poor farming family. He had very little education, he spoke with a strong Yorkshire accent, and he was a professional, non-commissioned sailor, not a gentleman-officer. Only his outstanding ability as a seaman, surveyor and navigator gained him the command of the expedition.

Cook had spent the early part of his career, before he joined the Royal Navy, as a merchant

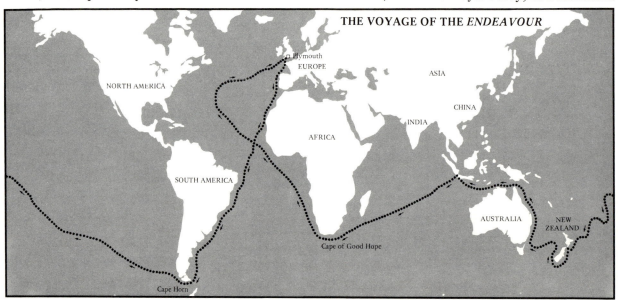

THE VOYAGE OF THE *ENDEAVOUR*

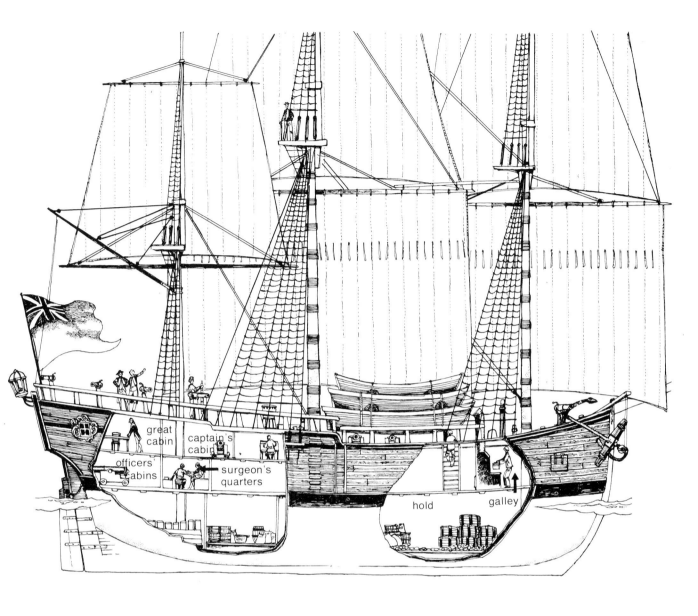

great cabin

captain's cabin

officers' cabins

surgeon's quarters

hold

galley

The Endeavour Bark started life as a coal ship, the Earl of Pembroke. She weighed 375 tonnes, and had a crew of 97 men. The exact cost to the Admiralty was £2,840. 10s. 11d.

She carried six four-pounder guns and eight swivel guns, and she sailed from Plymouth on 26th August 1768.

seaman on a collier, transporting coal from the north of England to London. The vessel chosen for the expedition, named the *Endeavour*, was of this north-country collier type. If Cook had been allowed to choose his own ship, he would certainly have chosen one just like her.

She was strongly built, with a rather flat bottom, a sturdy construction with plenty of storage space, but she was not very fast. She was about 30 metres long and 10 metres wide, not a large ship – 375 tonnes burthen – and she cost the Admiralty £2,800.

Though she was in excellent shape, much refitting was needed for a long ocean voyage. An extra sheathing of thin oak planks was added to her hull as protection against a greedy little creature called the teredo worm, which infests wooden ships in tropical seas. It was not enough, and on later voyages Cook had his ships sheathed with copper.

Endeavour carried six four-pounder guns on carriages, and eight swivel guns. She was fitted with new masts, but the work was delayed by a strike in the dockyards.

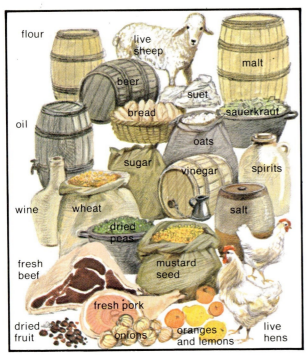

flour
live sheep
malt
beer
suet
bread
sauerkraut
oil
oats
sugar
vinegar
spirits
wine
wheat
salt
dried peas
fresh beef
mustard seed
fresh pork
dried fruit
onions
oranges and lemons
live hens

several others. One instrument he did not carry on the *Endeavour* was the chronometer perfected a few years before by another famous Yorkshireman, John Harrison. This instrument, really no more than a large, extremely accurate watch, finally solved the old problem of calculating longitude. The chronometer records the time at the place of departure, and a comparison between that time (registered by the chronometer) and 'local' time (found by astronomical observation) gives the longitude of the place.

Cook did carry a copy of Harrison's chronometer on later voyages, but his calculations of latitude and longitude were remarkably accurate on the *Endeavour* voyage even so. He was often not more than four or five kilometres out in his references for places in the South Pacific.

She was supplied for twelve months. (The voyage would take longer, but there would be opportunities to restock before then.) Among the most interesting items in the roomy hold of the *Endeavour* were a mysterious 'machine for sweetening foul water', green baize cloth for the great cabin (at special request of the Captain), 'portable soup' (something like primitive stock cubes), and various types of food including *sauerkraut* (pickled cabbage) which, it was hoped, would reduce the danger of scurvy. Cook was brilliantly successful at avoiding scurvy among his crews, but it is not certain that the *sauerkraut* had anything to do with it.

For trading with the natives of Pacific lands, the *Endeavour* carried iron nails (in great demand in Tahiti), mirrors, fish hooks, axes, scissors, coloured glass beads and – rather a surprise – children's dolls.

Cook made detailed requests to the Admiralty for scientific instruments, including a theodolite, a new, improved azimuth compass, a reflecting telescope like one he had found useful while making a survey of Newfoundland, and

Cook's own compass which hung from the ceiling of his cabin.

reflecting telescope

Right: The theodolite is a basic surveying instrument, which measures horizontal and vertical angles.

gifts and trading items
iron nails
axe heads
dolls
scissors
mirrors
glass beads
fish hooks

Apart from Cook there were three scientists on board: Charles Green, an astronomer; Joseph Banks, a naturalist; and Carl Solander, a botanist. There were also two artists, Alexander Buchan and Sydney Parkinson.

Tahiti to New Zealand

After several weeks in Tahiti, a place which seemed like a garden of Eden with its pleasant climate, friendly people and tropical flowers and fruit, Cook sailed to investigate the land discovered by Tasman over one hundred years before.

If we picture the *Endeavour* as she puts out to sea, we see not a stately square-rigger gliding in calm silence through a blue ocean, but a tubby little ship crowded with men, the captain arguing with the ship's scientists over who should use the green baize table, and instead of the slap of the waves and the gentle creak of the rigging, the air is full of the snorts of pigs, the bleating of sheep (for the best way to carry fresh meat is to carry the live animal), the squawk of Polynesian parrots and even the barking of a dog.

The weather began to grow worse, and the animals taken on board in Tahiti grew sick and had to be killed. The boatswain's mate died after drinking nearly a litre of neat rum. A comet was seen in the sky. On the rare calm days Joseph Banks, the young naturalist, put out a small boat and shot specimens of the sea birds near the ship. The anniversary of their departure from England was celebrated with a Cheshire cheese and a cask of ale.

Once or twice a fogbank or a low cloud was mistaken for land, but the heavy swell spoke of many miles of unbroken ocean. At the beginning of October the colour of the sea turned paler, often a sign of approaching land, and on the 6th, a boy at the masthead won the promised prize of a jar of rum when he was the first to cry, 'Land!'

It was the east coast of New Zealand's North Island. They put into Poverty Bay, named by Cook later because he could not get the supplies he wanted there. Tasman had been frightened off by the Maoris, who did not want visitors, and Cook hoped to avoid trouble if he could. He waited two days before sending a party ashore, and when the people showed signs of aggression, the Englishmen hastily retreated.

After several more attempts had failed, Cook's men were forced to fire in self-defence, and several Maoris were killed. Cook was sad. Unlike many people of his time he did not regard simpler races as 'savages', and he did not feel that he had the right to force them to do what he wanted. In Tahiti he had shaken his head in sorrow at the thought of the future destruction of the islanders' way of life, which their contact with Europeans was certain to cause. Few people would have realized that, or cared about it.

Fortunately Cook was later able to establish good relations with the people of New Zealand, who proved to be more friendly farther north. He admired them greatly, especially for their courage and high sense of honour. He reported in detail on their customs, and his account reads more like the work of a modern anthropologist than that of an eighteenth-century sea captain.

There were also, of course, some Maori customs that Cook did not like. He was shocked to find they made a practice of eating their conquered enemies.

He was intrigued by the close resemblance between the Maoris and the people of Tahiti (although the Maoris were far more warlike). He had a Tahitian with him who could speak quite easily with the Maoris. Cook did not know that, several centuries before, the ancestors of the Maoris had emigrated from Polynesia in their ocean-going canoes, and had settled in New Zealand.

Cook always regretted the deaths of the Maoris that his crew shot in self-defence

50

Maori warriors had elaborate facial tattoos, and the artists on board the *Endeavour* made many sketches of these warriors.

A Maori double war canoe. Similar canoes were used by the natives of Tahiti.

A decorated Maori war club, carved from bone and used mainly for ceremonial occasions.

The *Endeavour* was in New Zealand waters for nearly six months. During that time she sailed, as closely as wind and currents allowed, all along the coasts of both North and South Islands, passing through Cook Strait, which divides them, on the way south. Cook proved beyond doubt that New Zealand was not part of some undiscovered southern continent.

A Maori view of Captain Cook

When the first settlers came to New Zealand about eighty years after the *Endeavour* made her famous survey, an ancient Maori chief still remembered the coming of Captain Cook when he had been a boy. From Te Horeta we get a view of what European explorers looked like to the people of a newly discovered land, which makes an interesting change from the usual point of view – that of the Europeans.

The ship, said Te Horeta, seemed a thing of magic. Its crew were also thought to be supernatural creatures, goblins or demons. They propelled their boats while looking backwards (the Maoris used canoes and had never seen a rowing boat) which was most strange unless they had eyes in the backs of their heads. They had a magical way of killing a bird. They pointed a stick at it, there was a noise like the crack of thunder, and the bird fell.

Yet these were kindly demons. They gave food away freely. There was something hard like pumice stone, yet very sweet to taste, and there was something fatty, rather like the flesh of man but more salty (probably salted pork). One of the demons mysteriously collected shells and flowers. (This would have been Joseph Banks, or another of the scientists.)

The strangers invited the people on to their ship. Te Horeta went, with other boys, following the warriors who gave their cloaks in exchange for other goods. The one who was obviously the chief, the lord of the demons, handled the weapons and inspected the cloaks of the warriors with interest. He spoke little, but patted the heads of the children, who sat and watched, not daring to move about for fear of being bewitched. He gave some potatoes to the people, which they planted and tended carefully, becoming the first people in the country to eat potatoes. To Te Horeta, the lord of the demons gave an iron nail. He put the nail in his spear, and used it to make holes in the side boards of canoes. One day his canoe capsized, and the magic nail was lost. Te Horeta dived many times in an effort to retrieve it, but without success.

THE GREAT BARRIER REEF

Lookout Point
Cape Flattery
Cape Bedford
Cooktown ■
Endeavour Reef
Weary Bay
Cape Tribulation
AUSTRALIA

Just north of Cape Tribulation (so called 'because here began all our troubles') the *Endeavour* struck the reef. Cook wrote, 'The whole was cut away as if it had been done by the hand of Man with a blunt edge tool.'

coral

On the Great Barrier Reef

After completing his brilliant circumnavigation of New Zealand Cook sailed for the unknown eastern coast of Australia, and followed it all the way from Cape Howe to Cape York. For a long way he sailed the narrow channel between the Reef and the mainland, a bold and brilliant piece of navigation in a ship which, having no power of her own, was always to a large extent at the mercy of unpredictable winds and currents.

However even Cook did not complete this piece of sailing without accident. The voyage nearly ended on the Great Barrier Reef.

The *Endeavour* had sailed nearly 1,600 kilometres inside the Reef with a man constantly sounding the bottom and others in the yards and in boats watching for dangerous patches below the surface. Just after the leadsman had sounded 17 fathoms (30 metres), a safe enough depth, there was a horrible noise (like scraping a lamppost with a car) which brought Cook racing on deck. The *Endeavour* had grounded on a projecting spur of coral.

Immediately, all sails were lowered so the ship would not be blown farther on to the reef. As the tide went out she was left in four feet of water, and had she not been a stout, flat-bottomed collier she would have capsized. Cook hoped the tide would float her off again, but in spite of throwing 50 tonnes of ballast, including the ship's guns, overboard, she would not move. And she was beginning to leak badly. With all pumps going flat-out, and water still creeping up the hold, the outlook was nasty, but at the peak of the tide she suddenly lifted off.

The ship was clear of the coral, but she would quickly sink if the sea became rough. Cook decided to try 'fothering'. A spare sail was loaded with old bits of rope and other softish material, including sheep's dung as a kind of cement, and passed under the ship like a sling. The idea was that when the sail was pulled tight, it contents would be sucked into the hole and would block it, partly at least.

It seemed to work, and a few days later the ship was safely beached near what is now Cooktown. While the carpenters repaired the holes – they found the largest hole had been blocked by a large piece of coral which had broken off – the crew relaxed on shore, eating kangaroo and turtle meat and forcing down some kind of green plant that Cook, who knew its value against scurvy, forced them to eat.

Sydney Parkinson painted many plants and animals that had never before been seen by Europeans. His drawing of a kangaroo was used as a basis for this painting by George Stubbs in 1772.

After calling at New Guinea the *Endeavour* began her long voyage home, arriving at Dover in July 1771, nearly three years after she had left England. Cook was to revisit his discoveries on future voyages, but he always took two ships, as a safety precaution, remembering the time he had nearly lost everything on the Great Barrier Reef.

Lewis and Clark

To the West Coast of America 1803

When Thomas Jefferson became president in 1801 the United States was a very small country. Apart from the Indians only about four million people lived there. There were only thirteen states (now there are fifty) and these were strung along the east coast. People had settled in Ohio, Kentucky and Tennessee, and a few had crossed the Mississippi River. But the Mississippi was the real boundary of the country: the land beyond, twice as big as the land to the east, was mostly unexplored and unknown to the Americans of Washington, Boston and New York.

President Jefferson was determined to find out what kind of country lay to the west. It was not yet certain that this country would ever belong to the United States. A huge stretch up the middle, called Louisiana, was owned by France, and Britain had a claim to the Oregon territory in the north-west. But no one lived in these regions except a few trappers and, of course, the Indians. Jefferson saw a future United States stretching the whole width of the continent, from the Atlantic to the Pacific Ocean. In 1803 he bought Louisiana from the French for 15 million dollars, and that removed the main obstacle to his dream.

Thomas Jefferson's dream

If the United States government was to control the unexplored lands to the north-west, the next step was to explore them, to see exactly what Jefferson had bought in the Louisiana Purchase and, most important, to look for a route across the continent. As Jefferson knew, you cannot have power without access. If the Americans were ever to gain the lands of the West, they had to be able to reach them, to find a way to unite both coasts of the continent.

Long before the Louisiana Purchase, Jefferson had been making plans for an expedition to the west coast. He had decided its objective – the Columbia River, whose mouth had been discovered by an American ship in 1792. He had appointed as his private secretary a young man named Meriwether Lewis, who was not very well qualified as a secretary but had all the qualities Jefferson was looking for to lead an expedition of exploration.

The Lewis and Clark expedition, 'a darling project' as Lewis himself called it, was a military expedition. Lewis had been a regular army officer before Jefferson made him his secretary, and his friend and co-commander, William Clark, was also an army veteran. The ordinary members, about thirty in all, were privates.

The main purpose of the expedition was to find a good route to the Pacific, and in those days before railways or motor cars a good route meant a good water route. Starting from St Louis, which was then a little frontier town of a few hundred people, Lewis and Clark planned to follow the Missouri River to its source – something no explorer had done. From there they hoped to make a short overland connection with a river, probably the Columbia, which flowed to the west. They knew they must expect mountains, but they hoped to find the Missouri flowing in a pass through them. The Rockies were a bigger obstacle than they had bargained for.

Lewis and Clark discovered that the route Jefferson hoped for did not exist. That did not make their expedition a failure because it is

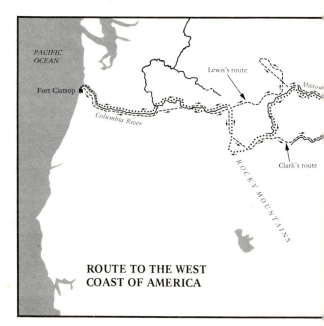

ROUTE TO THE WEST COAST OF AMERICA

Lewis and Clark trained their soldiers in St Louis during the winter of 1803

21 bales of gifts
powder and shot
swivel gun
air gun
tools
1,500 kg flour
50 kegs pork
340 kg salt
250 kg biscuit
1,000 kg parchmeal
275 kg grease
medicines
2 pirogues
17 metre keelboat

Mandan Villages
Fort Mandan
Mississippi River
Missouri River
St Louis

············ Outward route
---------- Return routes

almost as useful to know that a route does *not* exist as to discover where it goes.

Anyway, the expedition had other purposes. The chief of these was to make contact with the Indian nations of French Louisiana and tell them that the Great White Father now lived in Washington, not Paris, news that most Indians received very politely though without falling about in excitement or joy. However Lewis, and even more Clark, who in later years became the friend and adviser of the Indians, were also to try to establish peace between the different Indian peoples and to enforce order on the obstreperous Sioux who, as allies of the British during the American War of Independence, had often harassed American hunters and settlers.

The expedition had strictly commercial objectives too. Jefferson hoped that the valuable fur trade would be diverted from Canada to the American route once Lewis and Clark had proved that boats could navigate the whole length of the Missouri. Nor was science ignored. Before setting off Lewis took special lessons in botany and zoology so that he should know what to look for among the plants and animals in the unexplored West. The expedition sent back specimens of plants and skeletons of animals,

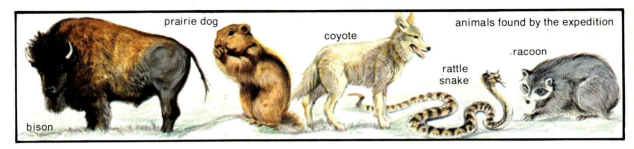

prairie dog

coyote

animals found by the expedition

rattle
snake

racoon

bison

and even some live creatures, including four magpies of an unusual kind. They also sent what Clark (whose spelling was terrible) called 'a liveing burrowing Squirel of the praries' – the animal now known as the prairie dog.

Along the Missouri River

From St Louis the expedition poled and paddled their way up river to the Mandan villages where they were to spend the winter of 1804–05. On the way they had several meetings with Sioux Indians, who were not always friendly. But at least no fighting broke out and Chief Black Buffalo became quite pleasant when Lewis allowed the squaws and children to look over his boat. It was a good deal bigger than the Indian dugout canoes, made from cottonwood trunks, which the expedition was to use on the higher reaches of the river.

Winter quarters were built among the Mandan villages, where the Indians were friendly. This place was often visited by French Canadian traders and hunters, and Lewis hired two of them as guides. More useful than either of the men was the wife of one of them. Her name was Sacajawea, or Bird Woman. She was a Shoshone (Snake) Indian who had been kidnapped and taken away from her homeland, 1,000 kilometres up the Missouri, several years before. When she joined the expedition Sacajawea was 16 years old and expecting a baby. Now, a pregnant teenager does not seem an ideal member of an expedition about to cross nearly 3,000 kilometres of unknown country. But Sacajawea did it, carrying her baby, who was born during the winter at Mandan. In fact she was one of the most valuable members of the expedition – an ideal ambassador when Lewis and Clark had to deal with Indians whom no white man had met before.

There was plenty to do during the winter. Local chiefs came to discuss matters with Lewis and Clark. One of them, Little Raven, brought a present of 25 kilograms of bison meat. Naturally, Little Raven did not carry that heavy load himself. He brought his wife along to do that.

wis and Clark spent the winter among the Mandan
dians. These Indians were farmers and traders,
d they also produced beautiful leather work and
ottery. There was a rumour that they spoke Welsh,
ut this turned out to be untrue.

inside a Mandan hut

These 'bull boats' were willow frames
covered with hide, and were light
enough for a woman to carry.

Lewis believed that his men would come to no harm from a grizzly— but he changed his mind.

Guns in 1805 had to be re-loaded after every shot.

Lewis and Clark ran a regular surgery, even carrying out simple operations like cutting off the frostbitten toes of a young hunter. (The winter at Mandan was as cold as in Siberia; Lewis recorded temperatures of −40°.) Hunting parties ranged up and down the frozen

river, towing the meat back on dog sledges, and a successful hunt was followed by that favourite recreation of the Indians, dancing.

Explorers seldom enter a country that is totally unknown. Usually, they have at least heard something about what they are likely to find, even though what they are told often turns out to be a long way from reality. Lewis and Clark were often finding things not quite as they had expected.

Grizzly bears, for instance. They were not the first white men to see a grizzly, though they believed they were, but they certainly saw more of these formidable animals than they would have wished. Their first 'yellow bear', as Lewis called it, was killed quite easily. He knew that the Indians feared the grizzly for its fierceness, and for the way it will hunt down its attacker in spite of the most severe wounds, but Lewis decided that the grizzly's reputation was exaggerated. Against a trained man with a gun, he said, they were not really dangerous.

A few days later he was not so sure. A grizzly was spotted in an open place 300 metres from the river. A party of six men set out to shoot him, and concealed themselves behind a rise 40 metres away. Four of them fired at the same time, and none missed: two shots, they discovered later, went through the bear's lungs. As if bitten by an annoying fly, the monstrous creature – well over two metres tall and weighing 275 kilograms – charged its attackers. The two remaining guns fired, and one of the shots broke the bear's shoulder, but he hardly hesitated. The guns of 1805 had to be reloaded after every shot, and there was no time for that as the grizzly bore down on the men. They took to their heels, the bear pursuing. Hiding among willows near the river they reloaded and fired again, but the shots seemed to do nothing except tell the bear where they were hiding. He came roaring down on two of them, who threw their guns away and leapt off a six-metre cliff into the river. The grizzly plunged after them, only a few metres behind. But at that moment, a shot from one of the others hit him in the head and killed him. The flesh proved too tough to eat, but they extracted several litres of oil from the carcass.

Lewis himself was later chased by a grizzly and took refuge in a tree, but the only serious hunting accident of the expedition happened when Lewis, dressed in brown leather after his uniform had worn out, was shot through the thigh by one of his men, who mistook him for an elk.

Indian moccasins gave little protection from these pears.

Indians hold the prickly pears in leaves to peel them.

Over the Rockies

Along the Missouri there was plenty of game, even if some of it was not easy to kill. (Lewis said he would rather fight two Indians than one grizzly.) But after they had left the Missouri and entered the Bitteroot Mountains, a part of the Rockies, food was scarce. Though helped by the Shoshone, Sacajawea's people, who lent them the horses they needed to get through the mountains, they had a hard time scrambling up rocky passes with no sign of a trail, and cutting a way through thick undergrowth. The spikes of the prickly pear went through their moccasins and hurt their feet, and the nights were getting cruelly cold, for they were nearly 300 metres above sea level and autumn was already with them. The Flathead Indians, who flattened the foreheads of their babies by pressing them with wooden boards, sold them more horses, and they crossed yet another ridge of mountains, through pine forest and over sharp rocky boulders. One

of the horses was killed for food, and another was lost when it slipped among loose stones and fell down the mountainside. There was no fresh water but there was plenty of snow to be melted. The lack of food was getting desperate when they reached the village of another great Indian nation, the Nez Percé. From there the expedition were able to launch their canoes in a stream which flowed into the Snake River and eventually into the Columbia. On November 7 Clark wrote in his diary, 'Great joy in the camp. We are in view of the Ocean.' He could just hear the noise of the waves.

The Lewis and Clark expedition was one of the best of all exploring expeditions: it was well planned and splendidly performed. But even Thomas Jefferson could make mistakes, and there was one very peculiar gap in his plans. He had not made firm arrangements to contact Lewis and Clark on the Pacific coast. Ships could and did sail frequently to that coast. Lewis and Clark had set out with the object of reaching the mouth of the Columbia, and they *had* reached it – and at about the time planned. But because no ship met them they had to spend the winter there, in a fort built of logs, and return overland the following year.

It was not a very enjoyable winter. For food they had dried fish, elk meat, birds of various kinds, and flour made by pounding the root of the arrowhead plant. It rained nearly every day. The local Indians were a miserable lot, corrupted by their contact with white traders. The Americans spent a lot of time making salt by boiling sea water. A stranded whale provided fat for candles. But in spite of these distractions, there was not much to do except watch the rain and wait for the spring.

fish

elk

arrowroot

birds

food for the winter at the coast

Return journey

On the journey back the expedition split into two at Travellers' Rest, their name for the place where they had spent a couple of relaxing days with the Flathead Indians before crossing the mountains. Clark turned south and followed the Yellowstone River to its junction with the Missouri. There he joined up again with Lewis, who had taken a route farther north.

During that time Lewis had a meeting with the warlike Blackfeet Indians. A party of Blackfeet, who seemed friendly, camped next to the American party. In the night they tried to steal their guns and horses. There was a fight and two Indians were killed – the only time that human blood was shed during the whole expedition. (One man died of a burst appendix,

which would have killed him wherever he had been.)

Lewis knew he could expect half the Blackfeet nation on his neck in rapid time unless he put a great deal of prairie behind him very quickly indeed. Fortunately, he and his companions had gained four Indian horses as a result of the fight while losing only one of theirs. They mounted at

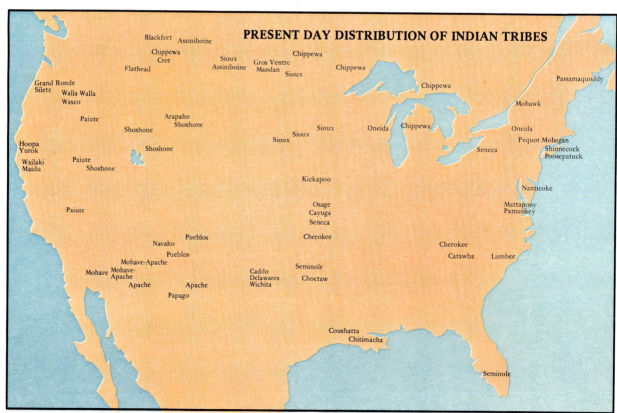

PRESENT DAY DISTRIBUTION OF INDIAN TRIBES

Sioux

Shoshone

Flat-head

Nez
percé

The Blackfeet Indians were a hostile tribe and could be very savage. They were usually to be found north of the River Missouri — as Lewis discovered. Clark, who took a southerly route had no trouble with the Indians.

once and rode for 100 kilometres without stopping. There was no sign of the Blackfeet.

In August Lewis's party arrived at the mouth of the Yellowstone and were soon reunited with Clark. A few days later they reached the Mandan villages again, where Sacajawea and her family left them. Her baby son was 18 months old and, according to Clark, was a beautiful boy. He had certainly had an exciting babyhood.

Lewis and Clark were greeted as heroes in Washington. In spite of the lack of a water route to the north-west, the expedition had been a success. Above all else, it had shown what a gloriously beautiful and rich country the Americans were to inherit, for after the Lewis and Clark expedition there was little doubt that the West would become part of the United States of America.

Burke and Wills

The Exploration of Australia 1860

Australia is the smallest continent in the world and the largest island. It was unknown to Europeans until Dutch sailors sighted the coast in the seventeenth century. In 1770 Captain Cook explored the eastern coast and gave New South Wales its name. The first Europeans to settle there arrived less than twenty years later. They were British convicts, sentenced to exile in an uncivilized colony instead of prison. They founded Sydney, capital of New South Wales.

Not until the early years of the nineteenth century did exploration of the interior of Australia begin. The Blue Mountains, which lie to the west of Sydney, were first crossed in 1813. Gradually, the coastal regions were explored, and in 1845 the explorer Sturt struck out from Adelaide into the centre of the continent. He discovered Cooper's Creek and the Stony Desert beyond.

A few years later gold was discovered in Australia, and colonists started pouring in by the thousand. New towns grew up in a few weeks. Melbourne turned from a frontier village into a large city. As the south-east became settled, people became more interested in the interior, where the great mass of the country lay untravelled and unknown, except to the Australian blacks – the Aborigines. Brave men set out into the waterless wasteland under a sun so hot that their thermometers exploded and the ink for their pens evaporated before they could write a word.

The greatest objective of these mid-19th century explorers was to find a route across the centre of the continent, from the south coast to the north. The first to succeed was the expedition of Burke and Wills, in 1860–61.

The preparations

The Burke and Wills expedition was the most elaborate and the most expensive that had ever been organized in Australia. The government of Victoria gave £6,000 for it, and the citizens of Melbourne gave £3,000.

At any time, Robert O'Hara Burke was a rather rash character. An Irishman, he had been in his time a soldier in Europe and a gold prospector in Australia. He had once fought a duel with sabres. When appointed to lead the expedition, he was police superintendent at Castlemaine. He had no experience of the Australian bush and he was nearly 40 years old. Though brave and energetic, he was a strange choice as leader.

Two dozen camels were bought in India and shipped to Melbourne with their Indian handlers. There were also 23 horses and several carts, one specially constructed to float across rivers. There were 12 tents, 20 camp beds, 80 pairs of boots (made by prisoners in the local gaol), books and instruments and cases for

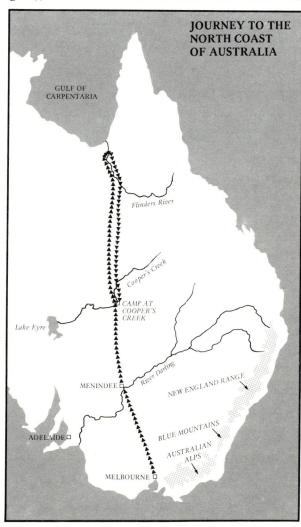

JOURNEY TO THE NORTH COAST OF AUSTRALIA

GULF OF CARPENTARIA

Flinders River

Cooper's Creek

CAMP AT COOPER'S CREEK

Lake Eyre

River Darling

MENINDEE

NEW ENGLAND RANGE

ADELAIDE

BLUE MOUNTAINS

AUSTRALIAN ALPS

MELBOURNE

24 camels

23 horses

12 tents

20 camp beds

The stores chosen by Burke weighed 21 tonnes.

several carts

80 pairs boots

camels' rum

books and cases

gifts for Aborigines

The expedition was given a rousing farewell when it left Melbourne on 20th August 1860.

specimens of plants and animals. There were flagons of lime juice to prevent the men getting scurvy (a disease resulting from a lack of Vitamin C), and 270 litres of rum to preserve the good condition of the camels. But Burke doubted if this were necessary, and when some of the men got drunk on the camels' rum, he got rid of the rest. The camels did not seem to miss their daily tot.

Stores and provisions were taken for eighteen months, and the total baggage, including beads and looking-glasses as gifts for the Aborigines, weighed 21 tonnes – more than one tonne per man. It was far too much. Burke had to sell some of his stores to the settlers before they plunged into the bush.

By the time they reached Menindee, the last settlement before the wilderness, some difficulties had already appeared. First, it was obvious that the expedition was too large, too slow and clumsy. Burke knew that soon after he had left Melbourne a rival expedition had left Adelaide with the same object: to reach the northern coast. He was therefore in a hurry – a dangerous condition for an explorer.

On the way to Menindee Burke quarrelled with several members of the expedition. His second-in-command resigned, and in his place Burke appointed William John Wills, who had been taken on as a surveyor to map the expedition's route. A careful, serious, 26-year-old Englishman, Wills was a very different character from Burke. But the two got on marvellously well. Their loyalty to each other in the terrible days ahead is one of the best parts of their story.

The expedition arrived at Menindee, just a few rough huts and a pub, in October 1860 – early spring in Australia. By the time it was ready to move again, the long hot summer was drawing perilously near. But Burke could not afford to wait until the autumn, in case the Adelaide expedition beat him to the north coast. He divided the expedition into two. An advance party of eight men would make for Cooper's Creek, his next objective, 650 kilometres north through uninhabited and mostly unknown country. The rest would stay at Menindee and bring up supplies later. A man named Wright was left in charge of the Menindee camp.

On to the north coast

Burke's streamlined party, with 16 camels and 15 horses, set off towards Cooper's Creek, where there was permanent water. (In Australia, the name creek was often given to a river which flowed only during the rains.) This was strange country, with birds, reptiles and insects which no one had seen before. There was a large caterpillar which made nests that hung from the trees, and an innocent-looking little black lizard that had a fierce and slightly poisonous bite.

The giant carpet snake looked worrying, but in fact it did not have a venomous bite.

More familiar animals could also be seen. Emus trotted off, with their heavy, unhurried tread, as the expedition approached. Tall kangaroos, on their ski-like feet, stood and watched the explorers pass. A carpet snake seven feet long gazed in amazement and was easily killed.

Wills, writing his diary on camel-back, found little to complain about. It was hard work loading and unloading the camels at each camp, and a diet of salted meat, and flat, home-made bread was not exciting. It was cold at nights but growing hotter and hotter every day. Although the ground was rocky, a fine dust arose and made breathing difficult. Some of the men wore veils over their faces to keep off the dust and the flies which pestered them. After 23 days, they reached Cooper's Creek.

Camp 65 at Cooper's Creek was made in the shade of a coolibah tree, with a stockade of young saplings to keep out unwelcome animals and Aborigines. The view was pleasant. Although no river flowed along the Creek, there were deep, wide waterholes, fringed with green reeds. Greeny-grey eucalyptus and gum trees lined the banks. Wild flowers grew, and flocks of brightly coloured birds arrived at evening, screaming wildly to each other as they kept an eye out for the circling eagles high above.

But Cooper's Creek was no paradise. Clouds of mosquitoes hummed around the animals and men at dusk. Always, in the bush, a man needed one hand free to beat off flying insects. There were also scorpions and centipedes six inches long, armed with powerful nippers. The rats were worst of all. They threatened to eat everything in the camp, and all food supplies had to be hung from the branches of trees to keep them safe. Sheer numbers made the Central Australian rat something more than a nuisance. At places along the route the camels had to tread very carefully to avoid damaging a leg in ground that the rats had threaded like a Swiss cheese.

The impetuous Burke decided to make straight for the north without waiting for Wright

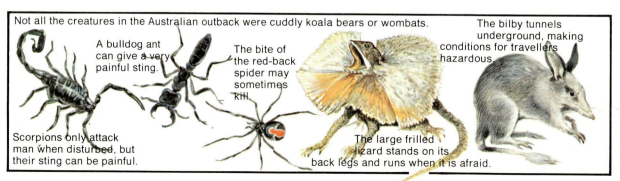

Not all the creatures in the Australian outback were cuddly koala bears or wombats.

A bulldog ant can give a very painful sting.

The bite of the red-back spider may sometimes kill.

The bilby tunnels underground, making conditions for travellers hazardous.

Scorpions only attack man when disturbed, but their sting can be painful.

The large frilled lizard stands on its back legs and runs when it is afraid.

Rats can burrow a metre deep, making the ground very dangerous.

When Burke and Wills reached Cooper's Creek, with its deep water holes, eucalyptus trees, colourful birds and plentiful fish, they thought it was a beautiful spot. But it was very hot – up to 42°C, and rats and mosquitoes were a constant nuisance.

and his party to bring fresh supplies up to Cooper's Creek from Menindee. He left four men at the camp at Cooper's Creek with William Brahe in charge. With him on the northward dash went Wills, John King, a young man who had been a soldier in British India, and Charlie Gray, a pleasant not too bright young fellow whom Burke had picked up along the way. They took supplies for three months: 150 kgs of flour, 45 kgs of bacon and salt pork, 45 kgs of dried horsemeat (they had killed two of the horses), about 15 kgs of biscuit, 20 kgs of rice, 20 kgs of sugar, 5 kgs of tea, 2 kgs of salt – these were the main items. They had guns and ammunition, but little spare clothing – and no tents. Each man had his bedroll or sleeping bag, and slept under the sky. The moon shone with a brilliance that some found disturbing.

Burke was too much of a gambler, too ready to take short cuts. His little party was travelling in midsummer – that was risky. They had no way of keeping in touch with their base – that was risky. They had no maps and no doctor – that was risky. Their rations were small for a journey across half a continent, and their lack of tents meant that they would have to spend some wet and miserable nights.

But they made it.

They passed through desert and through fertile plains, through woods and over rocky ridges. They struggled in gales and rainstorms, under a burning sun and under black clouds.

After nearly 1100 kms they struck the Flinders River as it flows towards the Gulf of Carpentaria. Soon they were floundering through salty marshes, in which Burke's horse twice got stuck and had to be dug out. When they noticed that the river rose and fell with the tide, they knew the sea was only a few miles away.

They turned back without seeing it. Men and animals were exhausted, the camels groaning at each step and Billy the horse weaker every day. Worse, the food supplies were running low. In the 57 days since leaving Cooper's Creek, they had eaten over two-thirds of their rations. Although they had found a plant, portulaca, which they boiled as a vegetable to provide them with Vitamin C, they had not been able to shoot many birds or animals to stretch out their supplies. When King did shoot a pheasant, it turned out all feathers and claws.

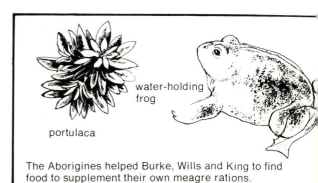

water-holding frog

portulaca

The Aborigines helped Burke, Wills and King to find food to supplement their own meagre rations.

The Burke and Wills expedition fought against all extremes of weather.

Eight hours too late

Somehow, they would have to get back to their depot at Cooper's Creek on much smaller rations. But they were already far weaker than when they had set out. Although there was no cause for despair, there was good reason to worry.

Charlie Gray did not feel well, but the others took little notice of his complaints. Burke had an attack of dysentery, a weakening disease probably caused by polluted water, but he seemed to recover. It rained all the time now. Each night they slept, or tried to sleep, in damp clothes. They all began to feel intensely weary – every movement became an effort. Wills thought this was the result of the hot and humid weather, but it was also a sign of malnutrition and exhaustion.

They made slow progress, and rations had to be cut down again. Wills caught Gray eating an extra portion of skilligolee (gruel made from flour and water) which he had stolen. When Burke was told, he lost his temper and beat Gray around the head. Gray was sicker than the others realised. A few days later they found him dead in his bed roll at dawn. It took them all day to dig a shallow grave to save his body from the dingos.

The horse Billy could go no farther and was shot, leaving just two camels. Poor Billy provided some meat, but Wills noticed there was not a scrap of fat anywhere on his carcass.

Parties of Aborigines, wondering what on earth these strange, gaunt, white fellows were up to, pestered them out of curiosity. They plodded on, only three of them now. What kept them going was the knowledge that they were not far from Cooper's Creek, where they would find plenty of food, new clothes, shelter, care and comfort. In spite of their stumbling weakness, so great was their eagerness to reach the camp that they covered the last thirty miles in one day's march.

Food was so scarce that an extra bowl of skilligolee caused a fight between Burke and Gray.

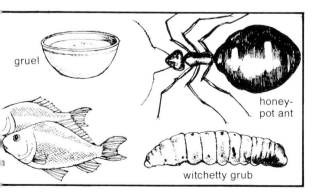

gruel

honey-pot ant

witchetty grub

Burke, Wills and King arrived back at Cooper's Creek only eight hours after Brahe's party had moved out.

Coo-ee! Coo-ee! The old call of the bushmen echoed through the trees, and the camp came in sight. But they heard no answering cry. When at last they staggered into the little stockade, they found it deserted.

Brahe had waited four months at Cooper's Creek. Wright had never managed to get there with the extra supplies from Menindee, and one of Brahe's men was ill and needed medical care. As Burke was so long overdue Brahe thought it likely either that he and his party were dead, or that they had returned by some different route, perhaps by sea or across Queensland. On 21 April, 1861, Brahe led his party out. On that same day, Burke, Wills and King arrived. They were too late by about eight hours.

Brahe had left some supplies behind: food for a month – but no clothes. They had a plan to follow the Creek to the west towards Mount Hopeless – how apt the name – where there were settlements. But the Creek soon split up into many small channels which petered out among the rocks. They dared not attempt a 240-kilometre march across the desert without water. The Aborigines, whom they had at first despised, helped them with presents of fish and showed them how to make nardoo – a cake of a kind made with the seeds of the nardoo plant.

A search party was organized at Melbourne. When Brahe appeared with the news that Burke was missing and long overdue, its task became suddenly urgent. It set out for Cooper's Creek, moving faster than Burke had done, and arrived on September 13. The Aborigines seemed excited to see the rescue party, and the rescuers soon found out why. Living among the Aborigines was a wreck of a white man, burned almost black by the sun, without an inch of flesh on his wasted body, and barely able to talk. It was John King.

He was the only survivor. Burke and Wills had died of starvation and exhaustion more than two months earlier. It was a poor reward for the first men to cross the Australian continent.

Grey are the plains where the emus pass
Silent and slow, with their staid demeanour;
Over the dead men's graves the grass
Maybe is waving a trifle greener.
Down in the world where men toil and spin
Dame Nature smiles as man's hand has taught her;
Only the dead men her smiles can win
In the great lone land by the Grey Gulf-water.

One of the last entries in Burke's diary showed how highly he thought of John King: 'King has behaved nobly and I hope he will be properly cared for.'

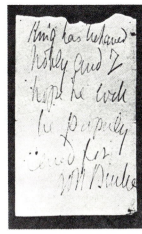

John King, the only survivor, was found five months after the return to Cooper's Creek, by a rescue party led by Alfred Howitt. It was thanks to the Aborigines that he was still alive. But he never fully recovered, and died of tuberculosis two years later, aged only 33.

Stanley

Down the River Zaire 1874

Most of the continent of Africa was a great blank to Europeans until the nineteenth century. Although regions near the coasts had been known for a long time Europeans had not, except in one or two places, travelled far into the interior.

The main pathways to the heart of Africa were its great rivers, and it was the courses of these rivers that became the targets for explorers from Britain, Germany, France and other European countries in the late eighteenth and nineteenth centuries. Mungo Park was the first to trace out the long curve of the Niger in West Africa. David Livingstone followed the Zambezi across south-eastern Africa to the coast. J. H. Speke discovered the main source of the mysterious Nile in Lake Victoria.

One great river remained, the Zaire. All that was known of the Zaire was the last stretch of about 150 kilometres before it poured into the Atlantic and stained the ocean brown for many miles. David Livingstone, during his lonely travels in

Central Africa, had stumbled on a great river called the Lualaba. We know that the Lualaba is in fact the upper Zaire, but Livingstone did not know this and he had an idea that it might even be connected to the Nile.

The secrets of the Zaire were finally unravelled a few years after Livingstone's death by Henry Morton Stanley.

Henry Morton Stanley

Like the river he explored, Stanley was a bit of a mystery. For a start his real name was not Stanley. He was born in Wales, and grew up in a workhouse from which he ran away to America when he was sixteen. In New Orleans he was adopted by a British businessman whose name, Henry Stanley, he took as his own. He added the Morton later because he liked the sound of it.

Stanley fought in the U.S. Civil War, on both sides at different times, and after other adventures, many of them rather shady, he became a reporter in the American west in the days of the wars against the Indians. At last he had found a

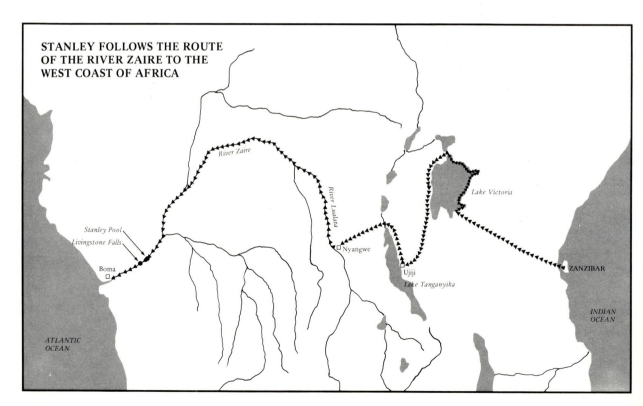

STANLEY FOLLOWS THE ROUTE OF THE RIVER ZAIRE TO THE WEST COAST OF AFRICA

River Zaire

River Lualaba

Stanley Pool
Livingstone Falls

Boma

□ Nyangwe

Ujiji
Lake Tanganyika

Lake Victoria

ZANZIBAR

ATLANTIC OCEAN

INDIAN OCEAN

job that suited his bold, inquiring, aggressive nature, and in 1869 the *New York Herald* hired him to lead an expedition into East Africa in search of Livingstone, who had vanished into the bush several years before. Two years later, after a spectacular march through East Africa, Stanley had his famous meeting with the great missionary-explorer ('Dr Livingstone, I presume?') on the shores of Lake Tanganyika.

Stanley's book about his search was a best-seller. He became famous, and he was able to persuade the *New York Herald* and the *Daily Telegraph* to put up £6,000 each towards his expedition of exploration into central Africa in 1874.

Preparations at Zanzibar

Stanley hired only three Europeans for his expedition, the two sons of a Kent boatbuilder named Pocock, and the clerk at his London hotel, a keen young man named Fred Barker. None of the three survived.

Frank Pocock, Fred Barker, a Zanzibari boy, Edward Pocock and Kalulu. None survived.

Zanzibar had become the usual starting point for expeditions to East Africa, and Stanley spent two months there arranging supplies and hiring men. It was a busy time. He had to choose the varieties of cloth, beads and wire likely to appeal as presents to the different people the expedition would meet on its way into the continent. Men had to be hired as porters, and Stanley was swamped with applications. At first it seemed that every sick or crippled man in Zanzibar wanted to go, followed by every crook and swindler. All had to be interviewed, and the unsuitable candidates weeded out. Even so, Stanley found later he had hired by mistake a man who had already murdered eight people. He was not an ideal member of an expedition into the wilderness.

Some unexpected problems also had to be dealt with. Stanley had ordered a portable boat, which could be divided into sections for carrying, but when it arrived it was too heavy. There was no craftsman at Zanzibar capable of modifying the boat, but Stanley managed to bribe a

African expeditions usually started from the island of Zanzibar. Stanley sailed from there to Bagamoyo.

carpenter from a passing mailship to stay behind and divide the *Lady Alice* (she was named after Stanley's fiancée) into lighter parts.

On 12 November a fleet of Arab dhows (sailing boats with large triangular sails) carried the expedition from Zanzibar to the mainland. It consisted of 224 people, six donkeys, five dogs, 72 bales of cloth, 36 bags of beads, four man-loads of brass wire, 14 boxes of assorted stores, 23 boxes of ammunition, two loads of photographic materials (not much use in the wilds), and fourteen other loads, including the medicine box, despatch boxes and cooking pots, and last but not least, the twelve parts of the *Lady Alice*.

mosquito

crocodile

python

On the Lualaba

The story of Stanley's expedition falls into two parts. During the first part, he marched to Lake Victoria. He explored the lake thoroughly by boat, and fought a battle against some people living on an island there. (Although Stanley never attacked first, he was a little too quick to answer any sign of aggression from Africans with a volley of rifle fire.) After visiting the kingdoms of Karagwe and Buganda, the expeditions arrived at Ujiji, on Lake Tanganyika. From there the second, more difficult stage of the expedition began – the march eastward to the Lualaba and down the Zaire to the sea.

During the first week or two the worst problem was not the aggressive cannibal tribes that Stanley had been warned against, but the thickness of the jungle. The trees began on the river bank like a black wall, and extended unbroken until lost in the haze around distant hills. It was a forbidding prospect. As Stanley's men plunged into the trees the sunlight vanished. Dew dripped from every leaf and trickled down every branch. Clothes were soaked, and Stanley's sun helmet felt heavy as lead.

They marched on through a damp and foggy twilight, bare toes scrabbling for a foothold, and loads wobbling dangerously on wet heads. The tidy, disciplined line which had marched so smartly across the dry and even plain turned into a ragged, untidy band slipping and sliding through the rain forest. By the time the end of the column passed the path had turned into a valley of thin mud, interrupted at frequent intervals by dank little streams ambling towards the main river.

It was a relief to embark the *Lady Alice*, and to slip down the river much faster and more comfortably than the men still struggling along the wooded bank.

The journey between Lake Tanganyika and the River Lualaba had to be made on foot, through dense jungle.

The *Lady Alice* originally separated into five parts. Later Stanley had it broken down still further, into twelve parts.

Stanley's boots, which he wore during the Zaire expedition. He had to do his own repairs to them.

Neither Stanley nor anyone else had any reliable idea of what they would meet or where they were going. Stanley still did not know whether the river would prove to be the upper part of the Nile, or the Niger, or, as was beginning to seem most likely, the Zaire. If it were the Zaire then they could expect trouble ahead, for they were about 500 metres above sea level, and rapids or cataracts were certain before they reached the sea.

Not all was gloom. On Christmas Day (1876) they took a holiday and held races on the river and on land. The most popular race was between ten young women of the expedition, for some of the men brought their wives with them.

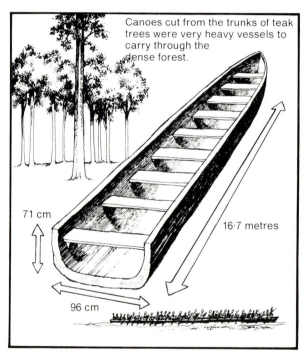

Canoes cut from the trunks of teak trees were very heavy vessels to carry through the dense forest.

71 cm

16·7 metres

96 cm

A week later they came to the first serious cataracts, later known as Stanley Falls. They floated down an empty canoe, to see what would happen, and watched in amazement as it was sucked into a whirlpool and reappeared, upside down, fifty metres downstream. Clearly the canoes could not pass the falls at this point, and they would have to be carried around the obstacle. That was easier said than done, for these were not lightweight canoes like the birchbark canoes of the North American Indians, but great heavy dugouts. They were hollowed out of the trunk of a teak tree more than three and a half metres in diameter, and they held forty or fifty people. To move them overland a roadway four and a half metres wide had to be cut through the forest and covered with

brushwood so the canoes could be hauled along it. To move them two or three kilometres by this method took several days, and the cruelly hard physical labour drained the strength out of the men.

The end of the journey

By the middle of March the expedition had reached a place on the Zaire where the river spread out into a large lake – Stanley Pool. A row of cliffs reminded Frank Pocock, the only European survivor besides Stanley himself – of the white cliffs of Dover. Everyone was feeling homesick. They had travelled over 11,000 kilometres since leaving Zanzibar. They were tired, and they had seen more than enough of the Zaire River.

Stanley knew that he had not far to go to the sea, and he knew also that he would soon reach a second set of cataracts. He might have left the river and struck through the forest, but having followed it so far he was determined not to be beaten by it. Exhausted by the mental strain of leadership as well as by physical hardship, Stanley no longer had the cool and level judgment he had two years before. His attempt to force the Livingstone Falls (his name for the rapids) was a mistake.

The Livingstone Falls are actually a series of about thirty rapids, which occur in two or three sections between Stanley Pool and the estuary of the Zaire. They bring the river down from nearly 300 metres to almost sea level, flowing part of the way through a narrow gorge. We can get some idea of the force of the river by

Although fruit was plentiful Stanley had to barter with natives for meat.

considering the rate of its flow: when it reaches the sea, the Zaire is flowing at the rate of about 50,000,000 litres *per second*. The current has been measured at 30 kilometres per hour.

Among these surging, boiling rapids, where black rocks thrust through the foam like giant teeth, a teak canoe 15 metres long was like a matchstick in a mountain stream. Two were lost in what Stanley called 'the cauldron' when he tried to float them down. Nine men were lost in several different accidents in one afternoon, although one turned up later after being washed

The jungle was so dense that a road had to be cut before the canoes could be carried round the rapids.

up on the opposite bank a kilometre down stream. Once again the body-breaking job of constructing a jungle road began, and the weakened men sweated among the trees and slithered over the wet rocks, while mosquitoes buzzed around their heads.

Stanley struggled on, making less than two kilometres a day, and sometimes calling a rest day for mending the canoes. Many members of the expedition were suffering from ulcers, especially on the feet, and from dysentery, a weakening disease like acute diarrhea. Although there was plenty of fruit in this part of the country, meat was scarce, and the expedition had almost run out of trade goods which could be exchanged for chickens or goats, or even dogs. When Stanley decided to be extravagant for once and buy a pig, it was the first meat he had tasted for four weeks.

On June 3, Stanley, whose hair had turned almost white during his crossing of Africa, was sitting on some rocks high above the river. Through field glasses he watched the approach of a canoe he had sent for. Suddenly, to his horror, it tilted and overturned. He could count eight heads in the water, bobbing around the overturned canoe. Not all those in the water managed to cling on, and Frank Pocock was one of those who were washed away. Six months before, young Pocock had asked his friend and leader, 'Do you really believe, in your innermost soul, that we shall succeed?'

Left: A variety of beautifully coloured butterflies can be found along the banks of the River Zaire.

Palla decius

Euphaedra zaddachi

Epitola crowleyi

Graphium tynderaeus

So many canoes were lost on the Livingstone Falls that many of the men still had to make their way round the falls on foot. Stanley himself lost his footing between two great boulders and fell eight metres – though he suffered nothing worse than bruised ribs.

'Yes,' Stanley had replied, 'I do believe that we shall all emerge into the light again some time.' Not all. Not Frank Pocock. Nor his brother Edward, lying four feet deep under an acacia tree in East Africa, nor Fred Barker, buried on the shores of Lake Victoria, nor Kalulu, an African boy whom Stanley had

they returned at the head of a column of porters carrying sacks of rice and sweet potatoes, bundles of dried fish and tobacco, a cask of rum and, a special treat for Stanley, a plum pudding and a jar of gooseberry jam. Some of the men were so hungry that they ate the fish and rice uncooked.

Spirits quickly revived. The pleasant experience of a full stomach banished despair, and suddenly everyone realised what he had somehow been unable to believe until this moment: they had reached the end of the journey. Before them the Zaire, quiet at last, spread out to embrace the Atlantic. They began to sing.

Kalulu was a young African boy whom Stanley adopted on his first expedition. He too died in the attempt to descend the Livingstone Falls.

adopted, drowned like Frank Pocock in the angry river. Nor several others, who had begun the journey but did not finish it.

The expedition was showing dangerous signs of breaking up. Stanley, whose Swahili nickname meant 'Breaker of Rocks', was losing heart. He was a determined man, the kind who does not give up, but he was as near defeat as at any time before in his adventurous career. The men were losing interest in life, finding no more comfort in smoking their cannabis, and in danger of dying from nothing more deadly than weakness and apathy. Many deserted, but wandered back again, having nowhere else to go.

At last, at the end of July, Stanley admitted that the river had defeated him when he announced that they would march overland to Boma, the town at the head of the estuary which would mark the end of their troubles. It was only five days journey away. The *Lady Alice* was left abandoned on some rocks by the river. (Meanwhile, the real 'lady Alice' had abandoned Stanley, and had married a rich young businessman who was not likely to disappear into the jungle for years at a time.)

Stanley sent two men ahead, and in a few days

They sang of the trials they had suffered, the pagans they had seen and the cannibals, and the waste places, and the meanness of people, and the great waterfalls. All joined in the chorus:

Then sing, O friends, sing; the journey is ended;
Sing aloud, O friends, sing to this great sea.

Nansen

Across the Arctic Ocean 1893

In 1879 an American naval officer, Lieutenant De Long, tried to reach the North Pole. In a motor yacht named the *Jeannette*, De Long sailed from San Francisco, stopped in Alaska to pick up some huskies, which he hoped to use to haul sledges across the ice, and headed north. The ship, like others before her which had ventured far into the Arctic Ocean, was soon caught in the fierce grip of the ice and frozen immovably. De Long hoped that the constant movement of currents, winds and temperature would eventually release her. But he was disappointed. For eighteen months the *Jeannette* remained imprisoned by the ice. Then she began to break up under the intense pressure, and De Long and his companions were forced to make a dash for land – the northern coast of Siberia.

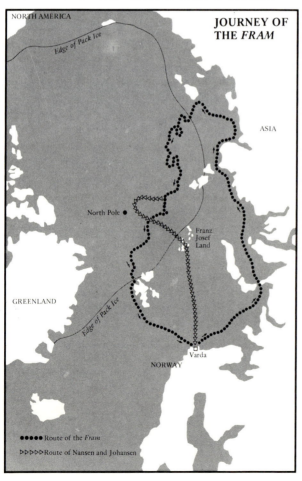

JOURNEY OF THE *FRAM*

NORTH AMERICA

Edge of Pack Ice

ASIA

North Pole ●

Franz Josef Land

GREENLAND

Edge of Pack Ice

□ Varda

NORWAY

●●●● Route of the *Fram*
▷▷▷▷ Route of Nansen and Johansen

Three years after the *Jeannette* was wrecked near the New Siberian Islands, a pair of oilskin trousers and one or two other items were picked up on the south-west tip of Greenland. They were identified as belonging to the *Jeannette*, and a meteorologist in Norway wrote an article in a morning newspaper in which he pointed out that the wreckage from the *Jeannette* must have drifted on an ice floe all the way across the Arctic Ocean and down the coast of Greenland.

The article was read by a young Norwegian scientist named Fridtjof Nansen. That day, the idea of the *Fram* was born.

Nansen had already led an expedition across Greenland, the first successful trip across the whole country. Stimulated by this success, Nansen started work on an idea for crossing the Arctic Ocean. The evidence of the *Jeannette*, together with other evidence of the ocean drift collected in Greenland by Nansen himself, showed that there must be a current flowing near the Siberian coast which crossed the Arctic Ocean, probably going quite near the North Pole, and flowed south through the Greenland Sea. The easiest way to get to, or near, the Pole, said Nansen, was to take advantage of this current and drift with the ice. Previously, all ships which had been frozen into the ice had perished, like the *Jeannette*. But Nansen believed a ship could be built which would 'ride' the ice without harm.

Nansen's plan was well received in Norway, but when he presented it to the Royal Geographical Society in London in 1892 the grey heads of British Arctic veterans shook in disapproval. Sir George Nares, who had led a recent Arctic expedition which set the record for farthest north, said that the basic rule for navigating in an icy region was 'stay near the coast'. Any ship that ventured out into the ocean would certainly be crushed. He did not believe in Nansen's theory of the Arctic drift. Likewise General Greely, who had led the last major American expedition towards the North Pole, gloomily said that he believed Nansen's plan could only lead to the deaths of all members of his expedition.

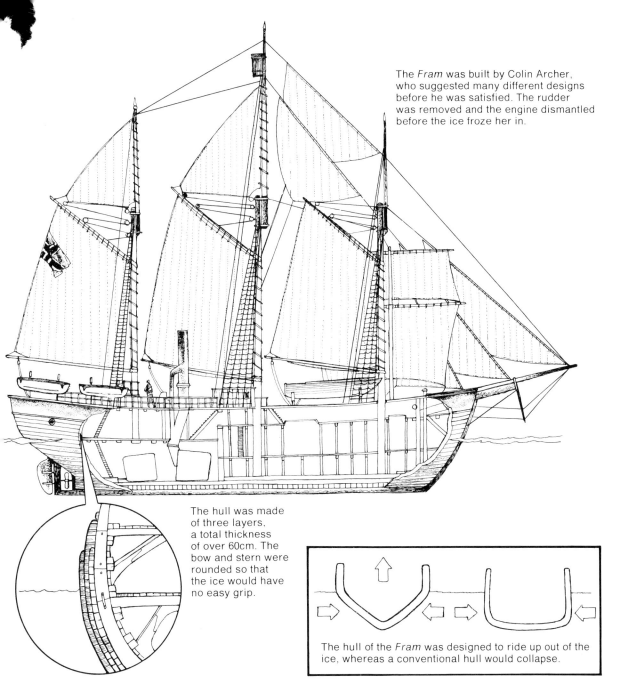

The *Fram* was built by Colin Archer, who suggested many different designs before he was satisfied. The rudder was removed and the engine dismantled before the ice froze her in.

The hull was made of three layers, a total thickness of over 60cm. The bow and stern were rounded so that the ice would have no easy grip.

The hull of the *Fram* was designed to ride up out of the ice, whereas a conventional hull would collapse.

But the experts, as experts sometimes are, were wrong. In 1893, together with twelve companions, Nansen set sail from Norway in the *Fram*, chugged through the loose ice of the Kara Sea, across the Laptev Sea, and turned north into the threatening ice. By 25 September 1893, the *Fram* was frozen in just north of latitude 78°, at almost exactly the point Nansen had calculated. The Norwegians settled down to wait.

The *Fram*

The *Fram* (the name means 'forward') was built by a famous Scottish shipbuilder, Colin Archer, and Nansen said that her success depended more on Archer's expertise than on his own ideas. Plan after plan of the ship was made, and model after

model. Fresh improvements were constantly being suggested, but in the end the ship was finished. She was not beautiful, because she was designed for strength, not speed or elegance. Her hull was shaped so that, when the ice closed in on her, it would push her upwards rather than squeezing her like a nut in nutcrackers. Thus the sides were rounded and the bottom almost flat. Her overall form was stubby – short and broad. Her hull had to be resilient as well as strong, and it was made of layers of oak and greenheart (from which fishing rods used to be made). It was over 60 cms thick altogether, and inside it looked like a hefty spider's web, with all the beams and braces supporting it. Bow and stern were rounded off to give the ice no easy grip.

carpentry

metal work

shoe making

sailmaking

Nansen's crew on the
Fram were kept busy.

The *Fram* was a little world in herself. The journey might take as long as three years, and she had to be equipped accordingly. There was a carpenter's workshop in the hold and a mechanical workshop in the engine room. A smithy, with anvil and furnace, was at first on deck and afterwards on the ice next to the ship. Finer metalwork, shoemaking (Nansen invented a special kind of Arctic boot, as well as the better-known Nansen burner and several other types of equipment), sailmaking and other work were done in the chart room or the saloon. There was even a windmill on deck which drove a generator to give some electric light in the long Arctic night, when the sun disappears below the horizon for several months.

On board the *Fram*

So carefully had Nansen planned his expedition that the main problem of the men on the *Fram* was boredom. Nansen felt quite guilty when he thought of the hideous ordeals of earlier explorers battling against the Arctic ice and compared them with his pleasant and peaceful life on the *Fram*. There were important scientific programmes to be carried out which meant daily tasks being performed and measuring instruments frequently consulted: the Norwegians discovered far more about the Arctic Ocean than had been known before. But after a time the scientific work became mostly mechanical, a simple routine to be performed every day like cleaning your teeth. There was plenty of time for writing and painting: Nansen kept a large diary and painted some attractive pictures of the 'icescapes' around.

Polar bears were also the cause of some activity. Because they have no natural enemies and do not know about the danger of men, polar bears have little fear, and it is sometimes necessary to shoot them just to make them go away. Nansen's men wanted them for meat and oil, and when a polar bear was sighted, everyone grabbed a gun and turned out on the ice. But the Arctic ice is not flat; it is a mass of lumps and ridges, and more than once a marauding polar bear disappeared among the hummocks, leaving behind the wreckage caused by its vigorous investigation of the ship's stores stacked on the ice.

Life on the *Fram* was not without other, more dangerous excitements. The big question was, would the ship stand up to the ice, or would she be crushed like all her predecessors, leaving the Norwegians with a rather slim chance of escaping to land?

When the *Fram* was drifting near the New Siberian Islands Nansen painted this water-colour – *The Polar Night, 24th November 1893.*

Inquisitive polar bears often visited the crew of the *Fram*, sometimes wreaking havoc with the stores on the ice. Sometimes it was necessary to shoot the bears as they simply would not go away.

Nansen had provisions for the time his crew would spend on the *Fram*, but fresh meat was also needed to supplement their diet.

When the *Fram* became locked in the ice there were a few anxious moments because no-one could be sure if she could stand the strain.

Early in 1895, well over a year after she had surrendered to the grip of the ice, the *Fram* came under intense strain as a ridge of broken ice, backed by hundreds of miles of steady, invincible pressure, advanced upon her. The air was full of the crunch and groan of the ice under stress, noises which had made early Arctic explorers think that strange monsters lived there. The timbers of the *Fram* creaked alarmingly in answer. The main pressure ridge built up and moved towards the ship just like a wave, except that it took not a few seconds to cover the distance, but several days.

The sturdy little *Fram* held out bravely, while her crew peered helplessly at the threatening ice ridge and kept their fingers crossed. At last the pressure relaxed, and with more creaking the *Fram* settled herself into a more comfortable position to continue her journey amid the encircling ice.

Attempt to reach the Pole

After some months had passed since the *Fram* had been frozen in, it became obvious that, although she was being carried across the Arctic Ocean, she would not pass as close to the North Pole as the Norwegians had hoped. Although this was a scientific expedition, concerned with studying the environment and not with breaking travel records, Nansen had naturally hoped he might be the first man to reach the North Pole. After a year and a half drifting slowly with the ice, Nansen decided to take one man, Hjalmar Johansen, and try to reach the Pole across the ice.

They set off with three sledges, drawn by dogs and loaded with kayaks, a type of canoe used by Eskimos. Nansen reckoned they could afford to travel fifty days before food for the dogs would run out – even if they killed the weaker dogs as food for the others on the way. But progress was slower than he had hoped. They ran into a patch of jagged ridges, and at every ridge, though most were only three or four feet high, the sledges had to be hauled and shoved over the obstacle by the two men. Not only did this slow them down, it was immensely hard and exhausting work.

Nansen and Johansen took with them dried fish and meat, butter, biscuits and one of Nansen's own inventions—a Nansen cooker.

A month after they had left the *Fram*, they were advancing only two or three kilometres a day, and Nansen decided they would have to turn back. Already they had travelled 250 kilometres father north than any man had been before, although the Pole was still over 320 kilometres distant.

The plan now was to make for Franz Josef Land, a remote place which had only been discovered a few years before. Although their watches had stopped, which meant that their calculations of their position were inaccurate, they reached land safely, and prepared to spend their third winter in the Arctic. The first two had been in the comfort of the *Fram*. Now they depended on their own devices.

They managed very well. They built a hut of stones, and laid in a store of walrus blubber – to burn for cooking and heating – and polar bear meat, which was their main food, although they still had some supplies from the ship. On this rich diet of bear meat, and with little exercise during the winter, Nansen actually put on a good deal of weight.

Although they had reached Franz Josef Land safely, perhaps the hardest part of their journey still lay ahead. In the spring they would have to make their way to Spitzbergen, the nearest place they were likely to find a ship. To cross a wide stretch of the Arctic Ocean in a couple of flimsy kayaks was not going to be easy. As things turned out, they never had to make that journey, which they had barely an even chance of surviving.

When spring came they started on the first stage of the journey, from the north of Franz Josef Land where they had spent the winter growing fat and bored, to the south. As they paddled their kayaks through the ice floes, they had several narrow escapes from disaster. Once, Nansen's kayak was attacked by an angry walrus – not a dangerous animal as a rule, though often, like this one, bad-tempered – and it managed to hole his kayak with its tusks before Nansen, swinging his paddle like a club and yelling furiously, beat it off. Later, they lashed the two

kayaks together to make a kind of catamaran with a sail. While they were resting on an ice floe, their 'boat' broke her moorings and drifted away. There was nothing to do but swim after it, and Nansen plunged into the water, which was hardly warm enough to melt the ice floating in it. He just made it to the two kayaks before he was paralysed by the cold. Only a very strong man could have done it, and without that extra layer of fat that he had put on during the winter, Nansen might have died in the icy water.

One morning, as Nansen was preparing breakfast in their camp, he heard a dog barking. At first he thought it must be imagination. Dogs meant men, but Franz Josef Land had been visited by men only three times in history. Nansen strapped on his skis and went to investigate. Soon he found dog tracks, then, quite distinctly, he heard a human voice. In the distance a figure appeared. Nansen waved his hat in the air, and the stranger waved back. A few minutes later he was face to face with a tweed-coated, pink-cheeked Englishman. 'Aren't you Nansen?'

'Yes I am.'

'By Jove! I am glad to see you.'

The stranger was a British explorer named Jackson, on an expedition to Franz Josef Land in his ship *Windward*, which was to take Nansen and Johansen home a few days later in far greater comfort that they could have hoped. Meanwhile, on the very day that Nansen had his amazingly fortunate meeting with Jackson, in another part of the Arctic the gallant little *Fram* burst out of the ice which had imprisoned her for so long. A few days later she steamed into a Norwegian port, where Nansen and Johansen were reunited with their shipmates.

Nansen lashed the two kayaks together to make a kind of catamaran.

Fuchs

To the South Pole 1957

Since the nineteenth century it has been the custom for scientists from all interested countries to club together every fifty years to study the polar regions. By 1950 many people had come to think that fifty years was too long an interval, and the Americans suggested holding an International Geophysical Year in 1957–58, halfway between the International Polar Years of 1932–33 and 1982–83. Geophysics means the physics of the Earth, and the idea of the IGY was to study the Earth – how it is made and how it works.

The purpose of the IGY was scientific investigation: it was not concerned with exploration. Yet some exploring was done, and of all the expeditions during the IGY much the most exciting was the Commonwealth Trans-Antarctic Expedition. It was led by Dr Vivian Fuchs and Sir Edmund Hillary. Hillary was a New Zealander, famous for climbing Mount Everest, and Fuchs was a British polar scientist. Other members of the expedition came from Australia, New Zealand and South Africa, which was then a member of the Commonwealth.

The easy route

The South Pole was first reached nearly half a century before the IGY by the Norwegian Amundsen, who got there one month ahead of a British expedition led by Captain Scott. Both Amundsen and Scott started their journeys to the pole from the Ross Sea and returned by the same route. This is the easiest way (or the least difficult!). Fuchs's plan was to start from the Weddell Sea, on the opposite side of Antarctica.

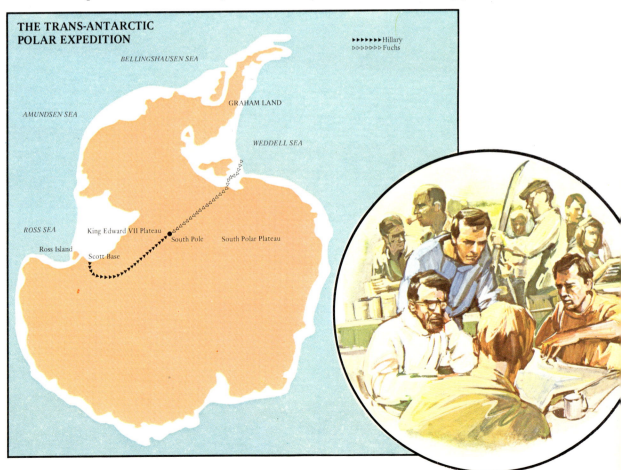

THE TRANS-ANTARCTIC POLAR EXPEDITION

▶▶▶▶▶▶ Hillary
▷▷▷▷▷▷ Fuchs

BELLINGSHAUSEN SEA

AMUNDSEN SEA

GRAHAM LAND

WEDDELL SEA

ROSS SEA

King Edward VII Plateau

South Pole

South Polar Plateau

Ross Island

Scott Base

Left: The expedition was planned and organized from small offices in London which always seemed to be overcrowded with both people and equipment.

Above: Even although the cabins of the Sno-Cats could be heated the members of the expedition preferred to keep the temperature below zero so that the snow and ice on their clothes and equipment did not melt.

Otter

Sledge

Sno-cat

The Fuchs expedition had many means of transport, both traditional and modern.

When he reached the South Pole, he would press on to Scott Base on the Ross Sea side. There was an American base at the South Pole where the expedition could enjoy a 'half-time' rest, and the route from there on had been covered by Hillary's party from Scott base, with supplies of fuel and food stashed along the way.

Much the hardest part of the journey was from the Weddell Sea up through ice-covered hills to the polar plateau. This region was completely unexplored. Fuchs's men discovered mountains which no one had seen before.

Besides travelling across the Antarctic continent, the Commonwealth expedition carried out scientific research. One of its tasks was to find out the thickness of the layer of ice which covers the land in Antarctica. By setting off a small explosion at the surface and measuring the sound waves echoing off the rock far below, the depth of ice could be measured. (In places it was nearly two kilometres thick.)

The seismic measuring instruments used in these tests were very sensitive. They would pick up the vibration of a motor vehicle many kilometres away, so everything had to stop while they were taking place. Before each test the geophysicist, Geoffrey Platt, rang a handbell to warn people to stop their motors and keep quiet.

The traditional way to travel in the polar regions is with sledges drawn by dogs. It is often the best way even nowadays, and Fuchs did make use of dog sledges. But his expedition was also equipped with everything that twentieth-century technology could provide. It had three types of aeroplane, from the little two-seater Auster, which could scout a route ahead for the vehicles, for the first 200 miles, to the much larger Otter, which was used to transport all the materials needed to build a forward base camp at South Ice. It had four types of motor vehicles, including four powerful American-made Sno-Cats, four Weasels (originally made as troop carriers during the Second World War), and several Ferguson tractors. All the vehicles needed careful maintenance and frequent repairs. It often seemed that the most important members of the expedition were the mechanics. But all the vehicles – the Sno-Cats especially – performed very well considering the conditions. Four of them chugged right across the continent.

Coping with the climate

The Antarctic is the coldest place on earth – much colder than the Arctic. The Fuchs expedition had to survive in temperatures almost incredibly low. During the Antarctic winter, one hundred degrees Fahrenheit of frost was measured at South Ice. The most bitter day of a British January would seem like warm summer in the Antarctic.

Travelling inside the vehicles, although not exactly comfortable when grinding over four-foot ice ridges, was warmer than skiing along behind the dog sledges. The Sno-Cats had double-glazed windscreens and heated cabins. But the cabins were never allowed to get too warm, or the snow brought in on boots would have melted, causing damp feet, which were more likely to be frostbitten. Polar travellers prefer temperatures below freezing because that keeps their clothes and equipment dry. Once snow and ice start melting, all kinds of problems arise. To climb into a sleeping bag that is frozen stiff as a plank may not sound comfortable, but it is better than a sleeping bag that is sodden with icy water.

Worse than low temperature is wind. A calm day at $-30°$ C is more comfortable than a windy day at $-15°$ C. Records of temperature do not give an accurate idea of how cold a person feels.

Seismic shots were fired to measure the depth of the ice. The sound waves echoed off the rock below.

Right: taking a gravity reading

Below: measuring the density of the snow

The Trans-Antarctic expedition was chiefly concerned with scientific investigation, not exploration.

The huskies were a traditional feature in an expedition that had many modern aids.

Everyone took a turn in the kitchen at Scott base. A favourite culinary trick was to add a measure of spearmint toothpaste when cooking peas!

Polar scientists talk about the 'chill factor', which takes wind strength as well as air temperature into account.

Intense cold has other unpleasant effects which experienced polar travellers try to avoid. All the members of the Commonwealth Expedition had a thorough check-up at the dentist before leaving. There is nothing like a few weeks in sub-zero temperatures to find out a weak tooth. Early polar explorers found that teeth could be cracked and splintered by the cold.

However cold weather, even the savage cold of the Antarctic, can be made harmless by careful preparation – and the right clothes. It is not necessary to wear layer upon layer of thick woollies and furs. More important is to keep a layer of warm air beneath, and the reason that polar travellers look so bulky in photographs is not only that their clothes are thick – they are also loose-fitting.

Whiteout

In March 1957 one of Fuchs's aeroplanes landed two men, with a sledge and ten days' food supplies, to make some geological surveys in the Whichaway Nunataks. (Nunataks are tops of mountains exposed above the ice.) They were only about thirty miles from the forward base of South Ice, so if bad weather made it impossible for the plane to pick them up they could, with reasonable luck, make their way to the base on foot, dragging their sledge.

Three days later Fuchs took off from Shackleton Base in the Otter, piloted by New Zealander Gordon Haslop, to pick up the sledge party in the Whichaways. But when they were in the air heading towards the nunataks, South Ice came through on the radio to say that the weather there was bad. The Otter turned back to wait for a better opportunity, hoping that the delay would not be lengthy.

The next three days brought no improvement. Fuchs, growing anxious, was in touch with South Ice every two hours. It was mid March, and the temperature was dropping daily. With winds of over thirty miles an hour and a temperature of -40° C, the sledge party would not be having an enjoyable time, and their chances of reaching South Ice safely did not look good. A constant, fast-moving drift, something like a sandstorm in the desert (though a good deal colder!) made navigation difficult. The sledge party might pass within fifty yards of the base camp without seeing it.

Two days later Fuchs and Haslop took off again. Fifteen metres above the ground it was as clear as a spring morning in the Mediterranean, but below that the drift was still sweeping across the land. Circling over the Whichaways, they saw no sign of their comrades and went on to South Ice.

approaching blizzard

snow drift

Antarctic scenery is very varied—from sastrugi, which makes travelling very difficult, to the aurora australis—the southern lights.

sastrugi—ridges of snow and ice

aurora australis

Fuchs and Haslop found that South Ice station w clearly visible from above, but when they landed they were in 'whiteout' conditions. Hal Lister fired a flare so that they would know where to cc

From above, the base at South Ice could be seen easily, but as they came into land, the ground disappeared and they plunged into a blinding whirl of snow. The aeroplane's skis touched down and, bouncing over the sastrugi (rough formations of ice fashioned by the wind), they came to a stop.

But where was the base? Somewhere close in front of them, they knew, but quite invisible. Fuchs got out and made his way gingerly forward, hoping to see the radio masts of the base. At every step he glanced back to make sure the plane was still in sight. The moment it disappeared he turned back. He knew that in whiteout conditions, just one step too far and a man may be hopelessly lost.

The depot in fine weather.

Antarctic bases can be quite complex affairs, with shelters for vehicles, dogs, scientific equipment and stores.

Next, Haslop tried to taxi the plane gently forwards. But the frozen skis refused to move. They unloaded the heavy barrels of fuel they had on board, and then the Otter pulled herself free and bumped her way forward – very slowly, for Haslop was afraid of running right into the buildings of the base.

A moment later a thickly clad figure emerged like some Abominable Snowman from the fog of snow. It was Hal Lister, one of the men from the South Ice camp, who had been guided by the sound of the Otter's engine. He shouted that he could not wait because his footprints would soon be covered and then he would not be able to find his way back to the base. So he and Fuchs ran back, along his footprints, talking as they went, until they found another South Ice man standing just within sight of the camp. Then all three ran back to the plane, adding a third set of footprints to the fast-disappearing track. Hastily they discussed plans before the two South Ice men ran back again.

Flying completely blind, Haslop got the Otter into the air and out of the drift. He flew back to Shackleton.

Several days later the weather at last cleared. The forlorn sledge party, who by that time had been out twelve days though only carrying food for ten, were picked up about twelve miles from South Ice. Both were suffering from frostbite, but not seriously. They were soon fully fit and ready to take part in the great journey across Antarctica.

Hidden dangers

A threat more frightening and more deadly than whiteout in the Antarctic is the crevasse. A crevasse is a crack in the ice layer which covers the continent, a split caused by the gigantic stress of shifting glaciers. Crevasses are usually found clustered together and bridged over with snow so that they cannot be detected. The first sign of their presence may be a horrible sinking feeling as sledge or vehicle sinks through a frail snow 'bridge'. Crevasses are often so deep that the bottom cannot be seen, and if a sledge or vehicle falls right in, it may be impossible to get it out.

In bad crevasse areas, the Commonwealth Antarctic Expedition sometimes tested for a safe route by sending a man in front to prod the snow with a long aluminium rod. If the rod still felt resistance six feet down, the ground was considered safe even for the 3-tonne Sno-Cats. In spite of all precautions, Fuchs and his men had several accidents. Three or four times they all but lost one of their vehicles in a deep blue Antarctic cavern.

Driving Sno-Cat 'Rock-n-Roll', Fuchs and his second-in-command, David Stratton, were making good progress when, without warning the 'ground' fell away and they were stranded over a hole 5 metres wide and at least 18 metres deep.

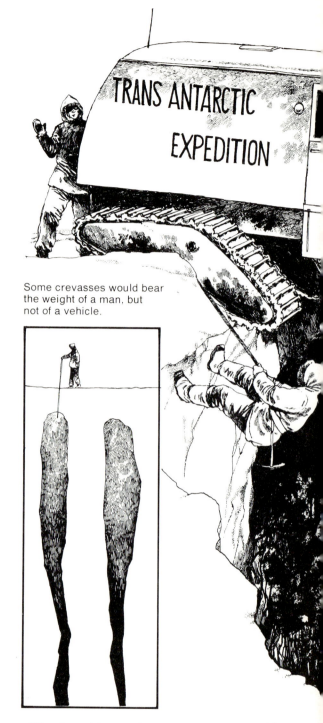

Some crevasses would bear the weight of a man, but not of a vehicle.

The first thing to do was to get out – and it was not easy. They were just able to reach one of the rear pontoons (the caterpillar-track 'wheels') and crawl along it, over yawning space, to safety.

The next step was to save the vehicle. First, everything that could be shifted was moved out of it. Then two other Sno-Cats were placed side by side behind 'Rock-n-Roll' and attached to her towing hook. Two Weasels were brought around to the front and tied by steel cable to the front

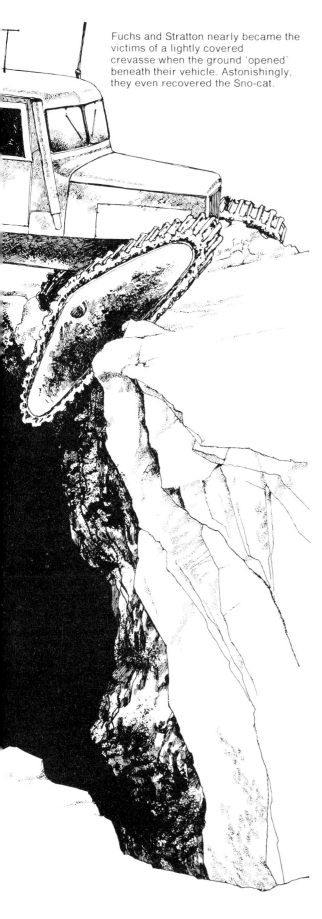

Fuchs and Stratton nearly became the victims of a lightly covered crevasse when the ground 'opened' beneath their vehicle. Astonishingly, they even recovered the Sno-cat.

axle of the stricken Sno-Cat. They were to act as an anchor, so that when the two Sno-Cats started to pull 'Rock-n-Roll' out backwards, she would not slip front first into the chasm. But it was not safe to start pulling until her front pontoon was in a position to slide out over the edge of the crevasse, and not catch on it.

David Stratton, brave fellow, was lowered into the crevasse on a rope. He cut a ledge in the ice which would guide the pontoon up out of the crevasse. Then a sixth vehicle, a big Canadian Muskeg tractor, pulled the pontoon into the right position.

All five rescue vehicles had to go into action at the same time. The Sno-Cats, using their emergency low gears, began hauling and, with a bump and a shudder, 'Rock-n-Roll' was drawn safely back from the edge of disaster.

The trans-Antarctic journey was probably the toughest that men and vehicles have made. Future travellers in Antarctica may find the remains of more than one of Fuchs's machines, deliberately abandoned. But the last part of the journey to the pole, across the plateau, was easier. Fuchs arrived at the South Pole, where Hillary waited to greet him, almost dead on schedule.

Together they travelled the second half of the journey, to Scott base. When Fuchs finally arrived there, he found a message from London to say that, like his New Zealand colleague, he had been made a knight.

Fuchs reached the South Pole on what he thought was 19th January. However Hillary, who was already there, was using New Zealand time, and by his reckoning it was 20th January.

Index